OUR
FINITE
MINERAL
RESOURCES

McGRAW-HILL EARTH SCIENCE PAPERBACK SERIES

Richard Ojakangas, Consulting Editor

Bird and Goodman: PLATE TECTONICS

Cowen: HISTORY OF LIFE

Kesler: OUR FINITE MINERAL RESOURCES

Matsch: NORTH AMERICA AND THE GREAT ICE AGE

Oakeshott: VOLCANOES AND EARTHQUAKES: GEOLOGIC VIOLENCE

Ojakangas and Darby: THE EARTH AND ITS HISTORY

OUR FINITE MINERAL RESOURCES

STEPHEN E. KESLER

Department of Geology
University of Toronto

McGRAW-HILL BOOK COMPANY

NEW YORK ST. LOUIS SAN FRANCISCO AUCKLAND DÜSSELDORF JOHANNESBURG
KUALA LUMPUR LONDON MEXICO MONTREAL NEW DELHI PANAMA
PARIS SÃO PAULO SINGAPORE SYDNEY TOKYO TORONTO

TO MY PARENTS

This book was set in Helvetica by Black Dot, Inc.
The editors were Robert H. Summersgill and Richard S. Laufer;
the designer was J. E. O'Connor;
The production supervisor was Dennis J. Conroy.
The maps were drawn by Andrew Mudyrk.
Kingsport Press, Inc., was printer and binder.

OUR FINITE MINERAL RESOURCES

Copyright © 1976 by McGraw-Hill, Inc. All rights reserved. Printed in the United States of America. No part of this publication may be reproduced, stored in a retrieval system, or transmitted, in any form or by any means, electronic, mechanical, photocopying, recording, or otherwise, without the prior written permission of the publisher.

1234567890 KPKP 798765

Library of Congress Cataloging in Publication Data
Kesler, Stephen E
    Our finite mineral resources.

    (McGraw-Hill earth science paper back series)
    Bibliography: p.
    1. Geology, Economic. 2. Mines and mineral resources. I. Title.
TN260.K47   553   74-23866
ISBN 0-07-034245-8

CONTENTS

Preface vii

ONE
WHAT, ME WORRY? 1

TWO
YOU CAN'T GET RICH QUICK—
  MINERAL EXPLORATION AND PRODUCTION 13

THREE
WATER—THE ABSOLUTE ESSENTIAL 23

FOUR
OUR COMMON MORTGAGE—
  SOME CONSTRUCTION MATERIALS 31

FIVE
LEGACY OF PAST OCEANS—
   OUR AGRICULTURAL AND CHEMICAL SUPPLIES   42

SIX
ENERGY RESOURCES—
   THE RISE OF THE ELECTRIC TOOTH BRUSH   52

SEVEN
THE FRAMEWORK OF OUR ECONOMY—
   STRUCTURAL METALS   66

EIGHT
EXTREME NATURAL CONCENTRATIONS—
   THE SCARCE METALS   78

NINE
MONEY AND DECORATIONS—
   PRECIOUS METALS AND GEMS   91

TEN
WHAT CAN WE GET FROM TODAY'S OCEANS?   99

ELEVEN
READ THIS   110

Appendix I. Important Ore Minerals   112

Appendix II. Production, Sources, and Prices of Mineral
   Resources of Importance in World Trade   115

Index   119

PREFACE

Only a century ago, the mineral resources of North America provided your ancestors with a ready avenue to affluence and prestige. The almost fantastic success attained by some who took this path can be gauged by the philanthropic activities of such wealthy families as the Carnegies, Guggenheims, Mellons, and Rockefellers. In spite of such achievements, however, only a few of our grandparents knew anything about mineral resources, and such knowledge ranked low on the list of qualifications for the "complete citizen."

Things have changed. Politicians and newscasters, as well as all the rest of us, are actively expressing opinions on whether the major oil companies have derived unfair profits from their operations. Pronouncements are available from everyone on whether metal mining desecrates the wilderness. Judgments can also be heard on topics such as strip mining of coal, disposal of nuclear wastes, arctic pipelines, and foreign expropriation. Clearly the resolutions to these controversial problems will have a profound effect on our life-styles and should be of concern to all of us. Very few, including some who are making their opinions heard, are sufficiently informed about our earth's mineral resources to make intelligent decisions in the proper context. For instance, if you are concerned with oil, you should know not only where it is found but how it forms, how it accumulates in the earth, how deeply we can drill to recover it and why. These are the factors that control our ultimate crude oil reserves. Because gasoline represents a major oil product, you also have to learn about platinum and lead, which, in effect, control the amount of energy we get from each barrel of oil and therefore affect the life of our oil reserves. Platinum, as you might learn, comes almost exclusively from one part of the world, which brings you to consider problems of international mineral supplies and foreign expropriation. Lead, on the other hand, is much more widely distributed, and you find yourself concerned with how lead deposits form, with methods of exploring for lead, methods of estimating how much ore is in the ground, and, finally, getting some feeling for how it might be mined and how these variables could affect the economic character of a lead deposit. Every mineral source, from asbestos to zinc, is woven into this complex web, and any balanced view of our mineral resources must take this into consideration. Information to provide this balanced view can be vexingly difficult to assemble for persons without training in mineral resources, however, because it is widely scattered and is usually not written for the layman.

This book is designed to help fill the information gap by providing a generalized discussion of the origin, distribution, and remaining supplies of our important mineral resources, as well as a consideration of some environmental, engineering, economic, and political factors that could affect their future availability. The intent here is to provide the sort of overview that is difficult to obtain without wide reading but is essential to consideration of any specific problem in the context of a complex industrial society. So, take a few hours and read the book from cover to cover. If your appetite is whetted, there are plenty of references you can consult for further information.

I am grateful to T. L. Kesler, J. A. Kesler, and R. W. Ojakangas for critical reading of the entire manuscript, and to A. J. Naldrett, J. C. Van Loon, and J. B. Currie for comments on specific chapters. In addition to persons acknowledged in the photograph captions, information concerning the illustrations was also supplied by G. O. Argall, Jr., D. H. Kupfer, J. Gittins, P. M. D. Bradshaw, D. R. Horn, M. Leahy, and M. Snedeker. Invaluable and efficient typing was supplied by J. Willott, M. Jurgeneit, and N. Stewart, and B. O'Donovan prepared many of the photographs.

Stephen E. Kesler

ONE

WHAT, ME WORRY?

INTRODUCTION
During the next 10 years, we will use more oil, aluminum, and other mineral resources than have been consumed since time began. As was forcefully indicated by the recent energy crisis, our supplies of these mineral resources are not infinite. Until recently, however, we have used the earth's mineral wealth to make our life more comfortable and hectic, without much thought of the needs of future generations. This was possible because world population grew slowly at first and our demand for mineral resources was limited. Within the last few centuries, however, our population has increased so dramatically (Fig. 1-1) that we are now consuming many of our mineral resources more rapidly than we can discover new sources.

This is only part of the story, however. Even more disturbing is the extremely rapid rate at which per capita use of mineral resources is increasing (Fig. 1-1). Americans on a per capita basis use over 4 times more steel, more than 100 times more aluminum, and literally 1 million times more uranium now than they used in 1900. Thus, even if we can stabilize our expanding population, technological advances will probably continue to increase our consumption of mineral resources. Furthermore, only about 10 to 20 percent of the population is really consuming the world's mineral

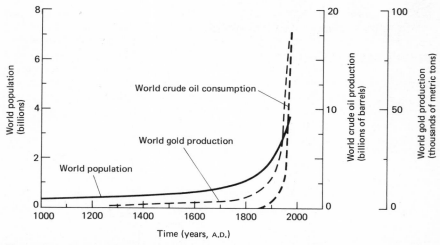

Figure 1-1 Relation between world population growth and the use of two widely different mineral commodities, gold and crude oil. Note that, although all three curves began to ascend steeply during the last century, the rate of gold and oil production increased more rapidly than the rate of population growth. This tremendously accelerated growth rate, which has been called the "deadly exponential," has placed enormous strains on our mineral industry to maintain adequate raw material supplies. (Data from U.S. Bureau of Mines Minerals Yearbook; U.N. Population Studies; and Park, 1968.)

production. Whereas the average North American uses about 4 gallons of oil each day, the average Asian uses almost none. With the Asian population over five times greater than that of North America, we can expect even greater strains on our mineral resources as underdeveloped populations reach for the fruits of our technological achievements.

Our situation is even more urgent when viewed in the context of earth history. It is generally concluded that the earth is about 4.5 billion years old and is divided into three concentric zones: a core, a mantle, and an outer skin—the crust—on which we live. During the last 3 billion years or so the earth has been slowly forming mineral deposits in the crust. Just over a billion years ago life began to develop, and civilized man arrived on the scene just a few thousand years ago. Thus, although we have been actively using the earth's storehouse of mineral resources for only the last 0.000001 percent of earth history, we are already well on our way to exhausting many of the reserves known to us.

Of course, the earth is still forming more mineral deposits, but even the most rapid ore-forming processes require thousands of years, whereas we can exhaust a large deposit in just a few decades. Thus, our mineral resources, with the exception of wisely used groundwater (Chap. 3), are finite and nonrenewable.

Is all this just a bad dream? After all, the world is pretty big. It would certainly look infinitely large if someone gave you a shovel and directions to dig up all the coal in Pennsylvania. Unfortunately, our mole's eye view just doesn't provide the proper perspective. Estimates made by highly qualified experts such as M. K. Hubbert of the U.S. Geological Survey suggest that remaining supplies of crude oil are adequate for less than one hundred years at expected future consumption rates. Many other mineral resources are similarly finite (Table 1-1).

What can be done? A commonly heard possibility is that we might recycle many of our mineral resources by recovering them from discarded items. This is a sensible goal and one toward which we have made considerable progress (Table 1-2). For instance, over 80 percent of the cars that are discarded annually in the United States are recycled to yield at least 10 million tons of metal. In contrast to cars, which consist of almost 90 percent metals, urban trash contains only about 10 percent. This amount is simply too low to be separated economically at present, and we are forced to throw away about 12 million tons of iron and steel and 1 million tons of other metals each year in the United States. To make matters worse, there are many mineral resources such as oil and potash that are completely decomposed when used and consequently offer no real possibility for recycling.

Another possible solution for impending shortages is the development of substitutes. This actually goes on all the time in industry. Aluminum can be used in pipes and wires in place of copper. Copper can take the place of silver in coins. Petroleum can be substituted for coal. However, substitutes serve only to shift the burden to another commodity.

It seems inescapable therefore, that we must find more mineral deposits if our civilization is to maintain or expand its present level of mineral consumption. Alternatively, we could extend the life of our mineral supplies

Table 1-1 Mineral resources estimated to remain in the United States in relation to expected demand for these resources in the United States

RESOURCES ADEQUATE FOR 10 YEARS	RESOURCES ADEQUATE FOR 10 TO 60 YEARS	RESOURCES ADEQUATE FOR MORE THAN 60 YEARS
Asbestos	Copper	Aluminum
Chromium	Gold	Barite
Fluorine	Lead	Clays
Mercury	Manganese	Gypsum
	Nickel	Iron
	Sand and gravel	Mica
	Silver	Molybdenum
	Tungsten	Phosphate
		Sulfur
		Thorium
		Titanium
		Uranium
		Vanadium
		Zinc

SOURCE: U.S. Geological Survey Prof. Paper 820.

Table 1-2 Amount of recycled material, as a percentage of total consumption, for some important metals in the United States (1971). The United States is one of the world's most efficient users of recycled materials.

METAL	PERCENT RECYCLED
Aluminum	16
Nickel	19
Zinc	22
Copper	22
Tin	26
Mercury	32
Lead	42
Iron	52
Antimony	60

SOURCE: U.S. Bureau of Mines Minerals Yearbook.

by decreasing consumption. This will occur automatically of course if production cannot keep pace with demand. However, attempts to limit mineral resource production artificially commonly run the risk of discouraging exploration for new deposits, something clearly undesirable in the context of growing mineral shortages.

It has probably occurred to you that we are not really exhausting the world's content of aluminum when we dump it into a trash heap. The amount of aluminum, iron, gold, and most other elements on earth has been essentially constant since the earth formed. Even petroleum, when burned, simply breaks down into water and carbon dioxide, which have not changed greatly in abundance in recent earth history. But, as you probably see by now, the important thing about a mineral resource is its high content of some desirable mineral commodity. For instance, there is gold in sea water and limestone, but there is no way to get it out without spending much more money than the gold recovered would be worth. Instead, we can recover mineral resources only from deposits where nature has concentrated them for us. It is these that we are exhausting and it is these on which we depend for our future supplies. What, then, is a mineral deposit? Can we estimate how many are left? Can we find more? That is the subject of this book.

WHAT CONSTITUTES A MINERAL DEPOSIT?

The term mineral deposit, as used here, refers to any natural concentration of mineral material that can be removed from the earth and sold at a profit. A mineral deposit is characterized by a combination of geologic, engineering, and economic factors. The geologic factor includes studies of the characteristics of mineral deposits as well as the processes that form them, all with the goal of finding more deposits. The engineering aspect includes ways of removing the mineral material from the ground and purifying it for use in industry. The economic phase, known as mineral economics, involves analysis of the many factors that govern availability, demand, and price of mineral products. Consideration of only one of these factors in estimating future mineral resources is like buying a new car on the basis of its color without looking at the price tag or motor. Thus, although we will emphasize the geologic features of mineral resources in this book, it is necessary to establish an appreciation for some of the important engineering and economic factors as well.

The Nature of Mineral Concentrations Think of a mineral deposit as some money that you accidentally dropped into a hole. If it is a lot of money, you will probably try to recover it. If it is only a few pennies, you might forget about it. Your decision is based on the amount or richness of the prize. The same is true in considering whether a mineral deposit is worth extracting. For petroleum and natural gas an estimate must be made of the amount and quality of fluid in the ground and how rapidly it can be pumped out. For solid mineral resources, we usually estimate the concentration of the valuable commodity in the rock (grade) and the total weight (tonnage) of rock that has this grade. For instance, a good gold deposit these days has at least 10 million tons of rock (usually called ore if it is profitable to mine) with a grade of about 0.2 ounce of gold per ton of ore. If gold were selling for $100 an ounce, a deposit of this size would contain at least $200 million worth of the metal.

This is not profit, however. It is necessary to spend tens of millions of dollars for mining equipment and many tens of millions more to actually extract the gold from its ore. A deposit of this sort could be expected to yield a profit, however. On the other hand, a deposit with 50 thousand tons each containing 0.2 ounce of gold, might not have enough gold in it to pay for assembling all the mining and milling equipment to recover the gold, and therefore it would not be an economic mineral deposit.

You might ask then, whether we could mine a gold deposit with 100 million tons containing 0.02 ounce of gold per ton. After all, wouldn't that be the same total amount of gold that was present in our minable deposit with 10 million tons of ore with 0.2 ounce of gold? In support of your suggestion you could point out that over the years technological achievements have permitted us to mine larger and lower grade deposits. For instance, in 1900, most U.S. copper mines were mining ore with about 2 percent copper. Today, the average grade is close to 0.5 percent and the tonnage of ore mined has increased dramatically. So, why not take one more step in that direction and mine huge tonnages of common rock for gold or copper? If this were possible, there would always be plenty of mineral resources and we wouldn't have anything to worry about, except perhaps falling in one of the many big holes that would be formed from all this mining.

The fallacy in this line of reasoning lies in the failure to recognize that most mineral deposits contain their element or commodity in some special, relatively easily recoverable form. Take zinc as an example. Almost all our major zinc deposits have grades of 3 to 10 percent zinc. This zinc is present in the mineral sphalerite, a zinc sulfide compound (Appendix 1), which contains over 60 percent zinc and which is closely intergrown with relatively worthless minerals containing no zinc. Sphalerite can be separated from worthless rock fairly easily and can be readily broken down to yield zinc metal. However, very little of the world's zinc is in sphalerite, and very little of the world's sphalerite is concentrated into ore deposits. Most zinc is found in very low background abundances in other minerals and rocks from which it cannot be recovered economically.

Fortunately for us, nature has locally concentrated some of these background amounts of zinc and other mineral commodities into deposits of minable grade and tonnage. In general, elements and commodities with relatively high background abundances, such as iron, aluminum, potash, and oil, require natural concentrations of only 5 to 50 times to form economic mineral deposits. In contrast, commodities such as gold, silver, and mercury, which have low background abundances, must have been naturally concentrated between 1,000 and 100,000 times to have formed ore deposits. In general, those elements or commodities that require less concentration, by nature, form larger and more abundant mineral deposits. This does not mean that only the highly concentrated commodities are in danger of depletion, however, because we use larger amounts of the less highly concentrated commodities (such as iron). Thus, nearly all mineral commodities are equal candidates for exhaustion.

Some Economic Considerations Many nongeological factors, such as price changes, technological developments, and government actions, have a strong control on the exploitation of known mineral deposits and the exploration for new ones. Price changes, whether gradual or abrupt, have a strong effect on the economics of many deposits. When the official price of gold was raised from $20.67 to $35 per ounce in 1934, previously uneconomic deposits were opened and world gold production rose almost 100 percent in less than 7 years. The recent increase in commercial gold prices to the $150-an-ounce range is sure to cause a similar jump in production.

For the past few years, price increases for most mineral commodities have out-paced inflation, thus allowing for more expensive exploration and exploitation in more remote areas. The Ertsberg mine (Fig. 1-2), at an elevation of over 12,000 feet in the highlands of New Guinea, resulted from such exploration. This 33-million-ton deposit averages 2.5 percent copper with significant additional amounts of gold and silver. The ore deposit was discovered in 1936 when two Dutch geologists noted a massive wall of copper minerals during a mountaineering expedition. Their account of the climb was rediscovered in 1959 and the Ertsberg was visited by American geologists whose overland trek to the area required 17 days. Political and economic considerations postponed development of the deposit until 1967, but by December 1972, when production began, about $150 million had been spent on the project for construction of mining and processing facilities, two towns, and a port. This sort of investment would clearly have been impossible and unneeded even 20 years ago when copper deposits were more readily discovered.

Incidentally, in case you were hoping we could stretch this idea of exploiting remote areas enough to permit the use of extraterrestrial mineral resources, forget it. There is little evidence of spectacular mineral concentra-

Figure 1-2 This photograph, taken from the area of the Ertsberg copper deposit in Irian Jaya (West Irian), Indonesia, shows the mile-long aerial tramways which carry ore from the mine in a steep descent from 11,750 feet elevation to the processing mill at 9,200 feet in the valley below. Concentrate from the processing mill is sent, in slurry form, through a 125-km pipeline to port facilities on the Tipoeka River on the south shore of West Irian. *(From Freeport Minerals Co., New York.)*

tions on the moon or elsewhere in space. Anyway, even an extreme example such as 500 pounds of gem quality diamonds brought back to earth from the moon would have a value of only $200 million, a small amount when compared to the $1 billion to $4 billion approximate cost of the mission. Besides, we produce only about 6,000 pounds of gem-quality diamonds annually on earth, and so your load might have a pretty bad effect on diamond prices.

The discovery of sizable new mineral resources can cause widespread price decreases in segments of the mineral industry. Large reserves of sulfur-bearing "sour" gas discovered in the 1950s and 1960s in western Canada had this effect on the sulfur market. This sulfur had to be removed before the gas could be marketed, and during the 1960s Canada became the second largest sulfur producer in the world. By 1968, world sulfur supplies exceeded demand, which brought about a large plunge in sulfur prices (Fig. 1-3) to a level at which many large sulfur accumulations in Louisiana, Texas, and Mexico (see Chap. 5) almost lost their status as ore deposits.

Technological developments of all types can have strong effects on the economics of mineral resources. For instance, before 1900, aluminum processing was very difficult and the metal was too expensive to be used widely. The development of new refining processes (Chap. 7) lowered prices and opened the door for aluminum to compete with other metals. Rock zones in Jamaica and Guyana, previously regarded as aluminum-rich curiosities, thus became minable deposits.

Developments involving the technology of scarce elements can cause enormous changes in demand. Since the 1940s, such developments have created an annual world demand for thousands of tons of uranium, an

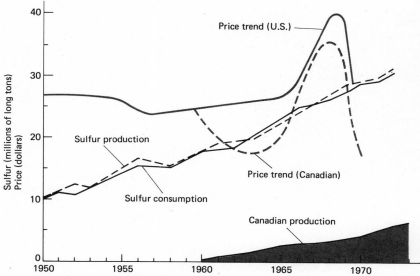

Figure 1-3 Variation in consumption, production, and price (Canadian and U.S.) of sulfur from 1950 to 1972. The most important feature of this figure is the fact that the price of sulfur fluctuated widely in the late 1960s and early 1970s although supply and demand were never greatly different. The pronounced drop in prices in the early 1970s can be attributed to the fact that a slump in sulfur demand by the fertilizer industry followed close on the heels of an expansion of production (including Canadian) to meet the strong sulfur demand of the middle 1960s. As you might imagine, the degree of error in any honest estimate of future sulfur demand or supply will be just about as large as the supply-demand difference that led to this price drop. *(Data from U.S. Bureau of Mines and Canadian Minerals Yearbooks.)*

element previously used only in small amounts as a yellow pigment and as a source of radium. In fact, there have been two uranium exploration booms, the first associated with accumulation of uranium for defense needs and the second, still underway, associated with our future energy requirements (Chap. 6).

Governmental actions are probably of greater importance to mineral deposit values than any other single area of economic activity. Governmental control is usually exercised through taxation and production controls. Taxation can take the form of an income tax, a tax on ore in the ground, a royalty on production, or any combination of these. In applying these revenue-generating devices, most progressive governments recognize that as a mineral deposit is extracted, a company is not only wearing out its equipment, it is also slowly putting itself out of business by depleting its major asset, the mineral deposit. This feature is unique to the mineral industry, and it merits, and sometimes receives, special treatment that will free money to be used in exploration for new deposits to replace those being exhausted.

In cases where the government controls the mineral deposits, a producer must also pay a royalty on production. Royalties on mineral production on public land in the United States yielded $490 million in 1973 for the federal government. In some areas, production quotas are imposed by governments, ostensibly to avoid excess production, which might result in decreasing prices and tax revenues. In some cases, such as the potash industry in Saskatchewan, the government even charges industry a sizable fee for the service of imposing these quotas. Import quotas, a related form of control, are also used by governments. Until mid-1973, the United States government limited imports of foreign oil to a relatively small percentage of total consumption. This was intended to stimulate development of domestic petroleum resources and limit U.S. reliance on the cheaper foreign oil.

Isolated government actions, such as boycotts and expropriations, can be of major importance to the mineral industry. Sometimes these actions produce unexpected results. A case in point is the United Nations economic boycott on the racially repressive regime in Rhodesia. Since this boycott was instituted, Rhodesia, which controls most of the world's high quality chrome ores (Chap. 7), has not been able to supply growing world demand for chromium, and alternative supplies have been sought. Somewhat ironically, two of the principal nations filling this supply gap are the USSR and South Africa, both of which have benefited from the large price increases growing out of the chrome-ore shortage.

The most important of isolated government actions is expropriation, the process by which a government can take over ownership (with or without compensation) of a business controlled by a foreign firm. The mineral industry is a prime target for expropriation, and many such actions, involving billions of dollars in resources, have taken place in Africa, South America, and Asia in the last few years. A sad example of this is the expropriation by the Chilean government of the El Teniente mine of Kennecott Copper Corporation, one of the world's largest copper mines. Although Kennecott transferred a 51 percent interest in the mine to the Chilean government in 1967, expropriation took place without compensation in 1971. One of the major claims made by Chile to justify their action was that Kennecott had derived excessive profits from their Chilean operations, a point strongly disputed by the mining company (Table 1-3).

A more compelling reason for many expropriations is illustrated in Fig. 1-4, which shows how much of the world's copper production originates in South America and Africa but finds its way to the United States, Europe, and

Table 1-3 Summary of earnings and distributions of the Kennecott Copper Corporation in Chile from 1916 to 1970

Value of metal sold	$3,430,000,000
Money expended in Chile (including $1,028,232,306 income tax 1930–1969)	$2,491,000,000
Money expended outside Chile	$430,000,000
Earnings to Kennecott	$509,000,000
Average earnings/year	$9,300,000

SOURCE: Kennecott Copper Corporation.

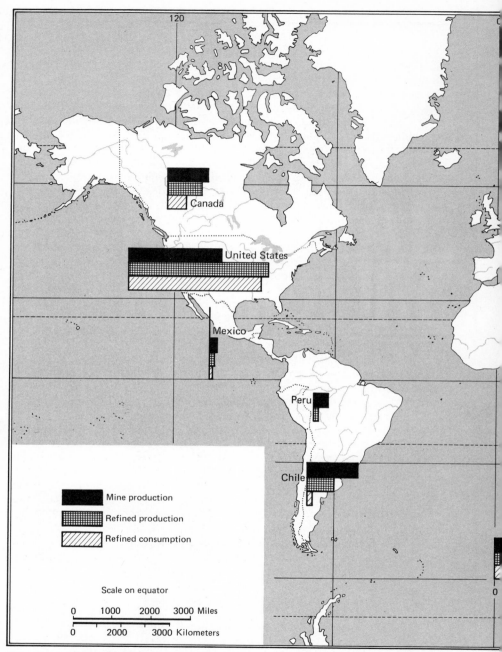

Figure 1-4 Mine production, refined production and refined consumption of copper in major copper-producing and -consuming countries for the period 1960 through 1970. Note that four major copper-mining nations, Peru, Chile, Zaire and Zambia consume almost no refined copper whereas some areas such as Europe and Japan,

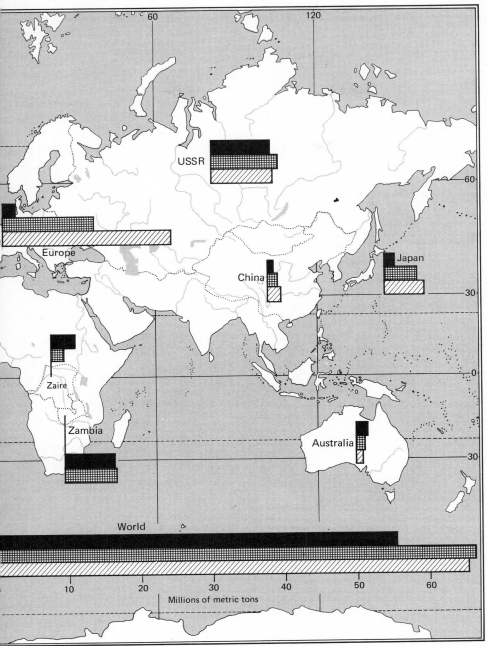

which mine almost no copper, fabricate imported copper into expensive products for re-export. This pattern of resource movement is true for most underdeveloped countries and is of major concern to their plans for long-term industrialization. *(Data from* Metal Statistics, *1971.)*

Japan for processing and fabrication. It then returns to the less industrialized countries in the form of more expensive transistor radios and cars. This pattern of movement is common for most mineral commodities, and it is of concern to underdeveloped nations because they see their mineral resources dwindling and no industrial base rising to take its place. Laudable as is the goal of national control of mineral resources, it cannot be used to justify illegal seizure, particularly when it is possible to purchase control of a company with a fraction of its future earnings. In view of the growing likelihood of uncompensated expropriation it is not surprising to learn that mineral exploration in many underdeveloped countries has stagnated, a situation that will further aggravate our near-term shortages.

TWO

YOU CAN'T GET RICH QUICK—
MINERAL EXPLORATION AND PRODUCTION

INTRODUCTION

There is a saying in the mineral industry that the best way to separate a man from his money is to give him a chance to invest it in mineral exploration. Everyone can conjure up visions of striking it rich by breaking into gold lodes or oil pools with a single swing of the pick. So, if given the opportunity to buy the pick for someone else in return for a share of the profits, most of us reach for our checkbook; the first time, that is. The second time someone offers us part interest in a patch of "moose pasture," we are usually more careful. The bitter truth is that only a very few of the properties on which exploration money is spent actually yield any profit (Table 2-1).

Mineral exploration is much like gambling. To be successful, you must have a system and enough money to withstand a string of losses. Experience in the Canadian mining industry indicates that an expenditure of $1 million over 3 years will give a company only a 3.2 percent chance of finding a mine. Thus, most modern exploration is carried out by large, well-capitalized companies.

Within the past few years, many federal and regional governments have also become involved in mineral exploration, ostensibly to stimulate exploration on projects too expensive for private exploration companies. For

Table 2-1 Summary of exploration activities of Cominco Ltd., a large Canadian mining company, which has pioneered in the evaluation and development of Arctic mineral resources

Number of deposits examined over 40-year period	1,000
Properties warranting major exploration expenditures	78
Properties on which ore bodies were found and mined	18
Properties yielding a profit from mining	7
40-year success rate	0.7%

SOURCE: A.T. Griffis, "Exploration: Changing Techniques and New Theories Find New Mines," *World Mining*, vol. 24, 1971.

instance, the Panarctic syndicate, in which the Canadian government purchased a 45 percent interest, carries out oil and natural gas exploration in the high Arctic of Canada where exploration costs are astronomical. If the Panarctic syndicate, which is made up largely of Canadian capital, had not been formed, exploration of Canada's Arctic islands might have been dominated by foreign firms having more risk capital than most Canadian firms. Thus, the indirect function of many government exploration organizations is maintenance of local control over mineral resources, a factor of growing political and economic significance and the same factor that motivates many expropriation efforts (Chap. 1).

Even if an organization is lucky enough to discover a truly economic mineral deposit, the odds are that it will be small. For instance, over 60 percent of all the gold produced so far in the United States has come from only 11 percent of the deposits. What's more, one of those deposits, the Homestake mine in Lead, South Dakota, accounts for about 10 percent of all the gold produced from all the deposits. On the basis of this historical perspective, even if an economic gold mine is discovered, there is far less than a 1 percent chance that it will be a rich producer. When these odds are stacked on top of the already bad odds of finding anything at all, it is little wonder that the grizzled prospector has faded from the scene.

Perhaps the ultimate indignity for the beleaguered prospector is the fact that even if an important deposit is partly exposed at the earth's surface, its true size and importance can easily be missed. One of the largest recently discovered gold mines in the world, in Nevada, recovers gold that is so fine grained that it is invisible to the naked eye and even to most microscopes. Some of the ground containing this deposit was sold for $2,000 by a prospector who could not see the gold. Usually, however, mineral deposits are covered by anything from 10 feet to 10,000 feet of barren rock, which makes exploration much more difficult and success more capricious. For instance, the Kidd Creek mine in Ontario (Fig. 2-1), which contains copper, zinc, lead, silver, and other metals with a market value of over $5 billion was covered by only about 50 feet of glacial drift and lake sediments. Nevertheless, a prospector lived and died in a cabin on the Kidd Creek deposit without ever realizing it was just below him.

Figure 2-1. This photograph of the Kidd Creek mine in Ontario shows the open-pit mine, which is nearing its maximum planned depth of about 760 feet. The deeper parts of the ore body (which continues to depths of at least 4,000 feet) will be mined by underground methods through a shaft that now extends to 3,050 feet below the surface. The 235-foot headframe for the shaft is just to the right of the open pit. Ore from the mine is shipped by rail for 17 miles to concentrating and zinc smelting facilities. *(From Texasgulf, Inc., Toronto.)*

EXPLORATION

In the good old days, many metal deposits and petroleum reservoirs were so easily discovered that the main requirements for successful mineral exploration were greed, perseverance, and luck. In most parts of the world, those days are gone, and we must look for less obvious mineral deposits with more modern methods. Modern mineral exploration usually involves prospecting a region for interesting spots and then carefully evaluating the most encouraging of these spots. Both chemical (called geochemical) and physical (called geophysical) methods can be used in prospecting. Evaluation of areas is usually done by drilling.

Geochemical Prospecting Almost all mineral deposits are surrounded by a zone that is enriched in the elements or compounds of the deposit. For instance, the La Brea tar pits, which smothered so many luckless sabre-toothed cats, were the messy surface leakage of the tremendous petroleum wealth of Southern California (Fig. 6-3). Although such zones are not usually so obvious to the eye, they can be detected by chemically analyzing natural materials. Thus, a copper deposit that is relatively near the surface might be decomposed by oxygen-rich rainwater trickling through the rock, much the same as a piece of iron rusts when left outside in the rain. Copper from the deposit will then concentrate in the soil around the deposit and will enter plants growing in the soil. Copper-rich water and soil will also be moved away from the deposit by streams (Fig. 2-2).

By collection of rock, soil, stream silt, stream water, or twigs from trees it is frequently possible to locate areas that appear to have unusually high amounts of the mineral resource of interest. Even more imaginative ge-

Figure 2-2 Copper content of silt from streams draining a previously unknown copper deposit in Puerto Rico. The copper reaches values of 0.1 percent in the silt near the deposit and decreases gradually downstream for almost 10 miles. Patterns like this have led to the discovery of many ore deposits. *(After Bergey, 1968.)*

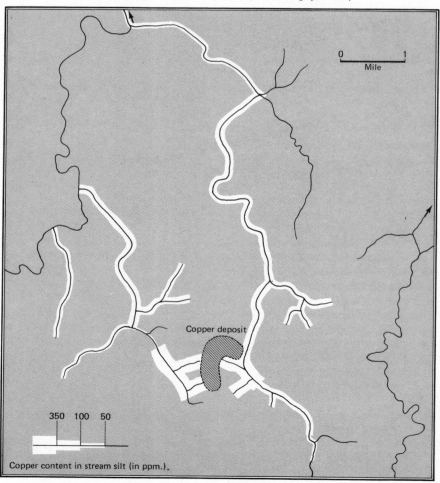

ochemical exploration efforts have been undertaken using air, snow, well water, glacial debris, and even the "blue haze," which consists of organic molecules discharged into the air by trees. Geochemical exploration can be credited with the discovery of ore bodies from Beltana, Australia to Rio Vivi, Puerto Rico including many deposits in the large Bathurst, New Brunswick and African Copper Belt districts (Chap. 8). Routine geochemical prospecting surveys usually cost between $1 and $100 per sample, depending largely on the accessibility of the area being surveyed and the density of sampling. A single company may collect several thousand samples per year.

Geophysical Prospecting Most geophysical prospecting involves measurements of physical properties, such as magnetic intensity, density, and electrical resistivity throughout a region. For instance, measurements of the earth's magnetic field, often from an aircraft (Fig. 2-3), can be used to locate potential iron deposits. The electrical conductivity of the metal sulfide minerals can be used to advantage in prospecting by measuring the resistivity or conductivity of large areas. Uranium and other elements can be sought with instruments that measure radioactivity. Important ore deposits discovered largely by geophysical methods include the Marmora iron deposit and the Kidd Creek massive sulfide deposit (Chap. 8) in Ontario and the Pyramid ore body at Pine Point, Northwest Territories (Chap. 8).

Figure 2-3 Many measurements useful in geophysical exploration can be made with airborne instruments. In this photograph, an instrument for measuring electromagnetic characteristics of the ground is suspended from a helicopter. The inset in the corner shows the instrument control panel in the aircraft. *(From Barringer Research, Ltd., Toronto.)*

Some mineral resources, such as petroleum and natural gas, tend to collect in underground structures that can be outlined by geophysical methods known as seismic surveys. These oil or natural gas traps (Chap. 6) can be delineated by sending shock waves through the rock and measuring the time it takes them to be reflected or refracted to other points on the surface. It is possible to use this method from ships as well as on land and, consequently, it is a major tool in evaluation of the continental shelf (Chap. 10). Because geophysical prospecting methods commonly require the use of more elaborate equipment in the field than do geochemical methods, they are correspondingly more expensive.

Evaluation Only a few of the many interesting spots indicated by geophysical and geochemical surveys qualify for detailed evaluation. This evaluation usually involves a thorough sampling program including collection of subsurface samples by drilling (Fig. 2-4). In many drilling programs, continuous samples can be cut or cored from the rock by using a hollow tube with diamond teeth at one end or by simply grinding the rock into small pieces which are flushed up to the surface by air, water, or a high-density water-mineral mixture called "drilling mud." Samples obtained by drilling are usually chemically analyzed or assayed, although in exploration for such

Figure 2-4 A typical core drill at work on a lithium-bearing pegmatite (Chap. 4) in North Carolina. The hollow pipes or "rods" between the two men are connected to a diamond coring bit that is rotated into the ground by the skid-mounted machine. Water, pumped through the rods by the hose (beside the top man), flushes out the pulverized rock leaving a cylinder or "core" of rock in the rods. *(From Thomas L. Kesler and* Mining Engineering.*)*

things as petroleum, water, brines, and natural gas, it is, of course, necessary to measure the amount and character of any fluid phase present in the drill hole.

Although some oil and gas wells reach depths of more than 5 miles, most nonpetroleum exploration drilling reaches depths of less than 1,000 feet. Such shallow drilling usually costs from $4 to $30 per foot of hole drilled, and even small mineral zones require several holes, often totalling 5,000 feet or more. Petroleum exploration usually involves larger drilling equipment and can be much more expensive. The average cost of one exploration hole off the east coast of Canada, for instance, was about $3.2 million in 1972.

PRODUCTION

If an exploration venture is blessed with success, it results in the discovery of a mineral deposit. As information on the deposit develops, other experts get into the act. Their concern is whether the new mineral deposit can actually be extracted and sold at a profit. In addition to evaluating the economic problems discussed in Chapter 1, it is necessary for them to determine the best method of extracting and processing the material. The economic character of any mineral resource is as strongly affected by these factors as it is by the odds that control its discovery. This is particularly true in the light of growing restrictions on environmental pollution by all phases of industry.

Extraction Mineral resources are recovered from the ground by either mining or pumping. There are two types of mines. Open-pit or open-cast mines consist of large excavations open to the sky. Underground mines consist of a subsurface network of tunnels and larger openings known as *stopes*, usually entered by a shaft. For ore bodies that are near the surface, it is usually cheapest to remove all the overlying waste (Fig. 2-5) whereas, for deep deposits, it is cheaper to extract the ore by way of a shaft sunk through the adjacent waste (Fig. 2-1). An often used rule of thumb is that total waste removed in an open-pit operation must be less than three times the volume of the ore, although larger amounts can be removed to recover high-grade deposits. This restriction usually limits open-pit mines to depths considerably less than 1,000 feet.

In addition to being safer, open-pit mining is usually cheaper than underground mining because it permits more of the ore in the ground to be recovered and larger amounts of material to be moved per day (Table 2-2). Thus, underground mines usually require higher grades of ore. It follows that many deposits that are economically minable at 200 feet could not be mined at a depth of 2,000 feet, a point of considerable importance to estimates of future mineral resources. In addition to the economic considerations, there are also physical restraints on the maximum depth of underground mining. The most important problems involve temperature increases with depth and natural explosion of rock in deep tunnels and shafts. In general, all factors limit mining to the upper 6,000 feet of the earth, although a few mines have reached depths of over 2 miles.

20 OUR FINITE MINERAL RESOURCES

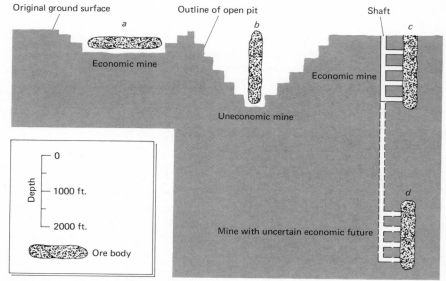

Figure 2-5 Open-pit and underground mines. If an ore body is entirely near surface (A) it can be mined by an open pit. The same ore body, turned on end (B) would require far too large an open pit to be economic but could be mined by underground methods (C). If the same ore body is deeper in the earth (D), it may well not be economically recoverable.

Pumping to remove fluid mineral resources can be carried out at depths greater than those of mining, but is still limited both by the difficulty of lifting liquids from great depth and by the fact that fluid-containing pores in the

Table 2-2 Approximate tonnages and grades of ore removed from some Canadian and U.S. copper mines. Note that most open-pit mines produce larger tonnages of lower-grade ore than most underground mines. Many of these mines, especially the open-pit mines, remove large tonnages of waste, also.

	OPEN PIT		UNDERGROUND		
MINE	TONNAGE (tons/day)	GRADE (percent copper)	MINE	TONNAGE (tons/day)	GRADE (percent copper)
Gaspé Copper, Que.	11,000	0.95	Louvem, Que.	850	2.76
Bethlehem, B.C.	16,300	0.43	Noranda, Que.	3,000	2.01
Battle Mt., Nev.	4,500	0.84	Anglo-Rouyn, Sask.	900	1.68
Sierrita, Ariz.	83,000	0.29	Granduc, B.C.	7,500	1.32
Morenci, Ariz.	59,000	0.83	White Pine, Mich.	23,000	1.00
Bingham, Utah	107,000	0.68	Magma, Ariz.	1,600	4.50
Chino, N.M.	23,000	0.86	San Manuel, Ariz.	63,000	0.69

SOURCE: *Canadian Minerals Yearbook* (1973) and *Mining Engineering* (1973).

rock tend to close as depth (and pressure) increase. These factors limit most effective pumping to the outer 20,000 feet of the earth, although some wells have reached depths of 28,000 feet. A novel application of technology to mining and pumping involves the use of major explosions to either increase the abundance of fluid-bearing openings in a rock or to shatter a solid ore body so that it can be leached by acids pumped down some wells and recovered through others. Several ore bodies have been shattered by conventional explosions and leached *in situ* and the United States government has experimented with nuclear blasts of this sort in a natural gas reservoir in Colorado. The presence of unusual radioactive silver in bullion produced in the USSR suggests that it was mined using nuclear blasts.

The maximum depths to which we can mine or pump material seem large to us, especially when descending 10,000 feet into the earth in a South African gold mine or watching a West Texas drilling rig pull 20,000 feet of pipe from a hole. In relation to the 4,000-mile radius of the earth or even the 10- to 30-mile thickness of the continents, however, we are not doing as well as a flea on a dog's back.

Processing After a mineral resource is extracted, it must still undergo a lot of expensive processing before it is ready for use in industry. At the simple end of the scale are mineral resources such as salt and coal which commonly require only cleaning, crushing, and sizing before sale.

At the other end of the scale are most metals and many nonmetals such as fluorine, which require a two-stage process. Zinc, for instance, often occurs in ore deposits in the form of an intimate mixture of sphalerite (Appendix I) and worthless minerals. This ore can be "beneficiated" by crushing and separating the waste to form a sphalerite concentrate. Whereas many zinc ores contain only about 5 percent zinc, the sphalerite concentrate can contain 60 percent zinc. This concentrate is then smelted, a high-temperature process in which the mineral sphalerite is chemically decomposed to yield sulfur and zinc metal (Chap. 8). Other mineral resources undergo processes that differ in specifics but have the same goal of complete or nearly complete reconstitution of the natural material. Crude oil, for instance, undergoes a complex distillation process called *refining* (Chap. 6).

There are all sorts of problems in mineral processing that might turn an attractive deposit into worthless rock. For instance, some of the high-grade lead-zinc-copper sulfide ores of the Bathurst area in New Brunswick are so fine-grained that they cannot be beneficiated adequately enough for smelting. Sometimes, the presence of an undesirable element, such as arsenic in a copper sulfide concentrate, will cause so much of a problem in the smelting process that the concentrate simply cannot be smelted economically and the deposit remains dormant. Finally, most mineral processing yields large amounts of waste (Table 2-3) that must be disposed of in accordance with growing restrictions on environmental pollution.

Table 2-3 Example of the volumes of waste generated by the mining and processing of enough rock to produce 1 ton of copper metal from a copper deposit originally containing 0.6 percent copper. First the copper ore is mined, then the ore is concentrated, and finally it is smelted.

PROCESS	VALUABLE PRODUCT (approx.)	WASTE PRODUCT (approx.)
Mining	150 tons of ore	300 tons of waste
Beneficiation	3 tons of concentrate	147 tons of tailings
Smelting and refining	1 ton of copper	2 tons of sulfur and iron (as SO_2 and slag)

SOURCE: *The Earth and Human Affairs*, 1972, Fig. 1-7.

CONCLUSIONS

With complications such as these governing the successful exploitation of a single mineral deposit, it is easy to see why predicting future mineral supplies is an uncertain business. The degree of uncertainty in such estimates can vary from today's deposits and economic conditions for which there is usually reliable, complete information to the distant future's expected discoveries and economic developments. In most estimates of remaining mineral supplies, it is common to reflect this uncertainty by using the term *reserves* for known deposits that are exploitable under present conditions, and the term *resources* for low-grade deposits that might become economic in the future as well as undiscovered deposits in areas considered favorable for exploration. We will use these two terms in this way in the following chapters.

THREE

WATER—
THE ABSOLUTE ESSENTIAL

INTRODUCTION

Importance of Water Water is, without doubt, the most important of all mineral resources. Unlike almost any other mineral resource, it is used by all people, regardless of their status in our modern economic system, and without it, no civilization can persist. Even in primitive societies, where individual water use is less than 1 gallon per day, water use for agriculture can be significant. For instance, over 50 gallons of water are required by one corn plant to reach maturity. A pound of wheat requires 60 gallons of water and a pound of rice necessitates over 200 gallons.

In more technologically advanced countries, per capita water use can exceed several thousand gallons per day. This amount includes all water removed by man for household uses, manufacturing, and irrigation. As can be seen in Table 3-1, irrigation and thermoelectric power generation far exceed all other uses in importance. Thus, Idaho residents used about 22,000 gallons per person per day in 1970 whereas each Vermonter used only about 250 gallons per day.

In addition to its importance in other sectors of the economy, water is of overriding importance to the mineral industries. Most methods for proces-

Table 3-1 Water withdrawal and consumption in the United States (1950 to 1970) in billions of gallons per day

	1950	1960	1970
Population (millions)	150.7	179.3	203.2
Public supplies	14	21	27
Rural, domestic, and livestock	3.6	3.6	4.5
Irrigation	110	110	130
Thermoelectric power	40	100	170
Other self-supplied industries	37	38	47
TOTAL	200	270	370
Fresh groundwater	34	50	68
Saline groundwater	—	0.38	1
Fresh surface water	160	190	250
Saline surface water	10	31	53
Reclaimed sewage	—	0.1	0.5
Water consumed by off-channel uses	—	61	87
Water used for hydroelectric power	1,100	2,000	2,800

SOURCE: Murray and Reeves, 1972.

sing mineral resources require large amounts of water. In the late 1800s and early 1900s rich precious metal mining areas in western North America, such as the Comstock lode in Nevada, were supplied with water by large water transfer systems. The Kalgoorlie gold mining area of western Australia depends on a 400-mile waterline from the coast; several rich nickel deposits in this area cannot be mined until more water is available. Prevention of pollution is a growing factor of importance in the disposal of waters used by the mineral industries. This is accomplished both by purifying the water before releasing it into the local surface water system or by reusing the water in the process. At Butte, Montana, for instance, one of the major copper-producing areas in North America, beneficiation and smelting require about 100 million gallons of water daily, over half of which is reclaimed, purified, and recycled.

Hydrological Cycle Of the 359 quintillion (10^{18}) gallons of water estimated to be on the earth, a remarkable 97.2 percent is in the ocean. Ice caps and glaciers account for most of the remaining water leaving about 0.65 percent of the total free water to be found in lakes and rivers, underground reservoirs (groundwater), and the atmosphere. Furthermore, streams and rivers are far smaller (0.02 percent) water reservoirs than is groundwater (0.63 percent). There is a constant exchange of water between all these reservoirs at rates controlled by climate, with precipitation being the main method of replenishing the freshwater supply. Thus, areas of present and potential water shortage in North American are those receiving less than about 20 inches of precipitation annually (Fig. 3-1). This includes northern Mexico, much of the western United States, the Canadian prairies, and most of the Arctic.

Most agricultural, industrial, and domestic uses for water require that it be relatively free of dissolved material. For this reason, sea water, which contains about 3.5 percent dissolved solids (Table 3-2) is not usable without expensive treatment. This treatment, known as desalinization, is generally too costly to be widely used except in very arid areas where no other options are available. Similarly, suggestions to use both melted icebergs and water

Figure 3-1 Distribution of areas in North America that receive less than 20 inches of precipitation annually (unshaded). Surface subsidence due to groundwater withdrawal is widespread in this area and some regions, such as the Llano Estacado, are rapidly depleting their groundwater supplies. The NAWAPA water-transfer system is one of several proposed projects by which future shortages might be alleviated.

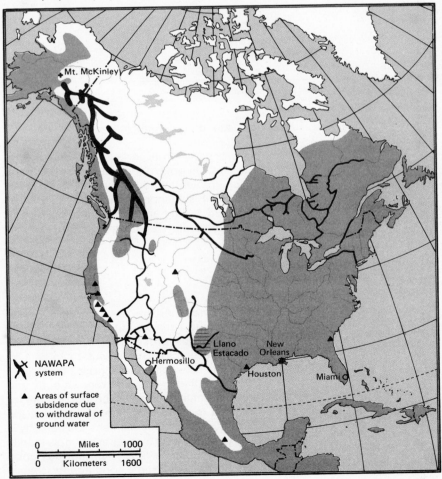

Table 3-2 Approximate composition of average river water and sea water. Note that the high dissolved solid content of sea water is largely sodium chloride (salt). All data are expressed in micrograms per milliliter (µg/ml).

	AVERAGE RIVER WATER	SEA WATER
Sodium	6	10,800
Magnesium	4	1,290
Sulfate	17	2,700
Chloride	8	19,400
Potassium	2	392
Calcium	15	411
Manganese	0.07	0.002
Iron	0.7	0.003
Copper	0.007	0.002
Bromine	0.020	67
Iodine	0.007	0.064
Gold	0.000002	0.000011
Lead	0.003	0.00003
TOTAL DISSOLVED SOLIDS	120	35,000

SOURCE: Skinner and Turekian (1973, Table 6-1).

condensed from the atmosphere are of potential interest only in local extreme situations.

The bulk of our water, therefore, is drawn from surface and groundwater reservoirs, and it is these amounts that will govern the water supply in most areas. Surface water provides a much greater fraction of our worldwide water needs than does groundwater because it is cheaper. Although groundwater requires more expensive investments in drilling and pumping equipment, it has the advantage of more constant temperature and composition as well as freedom from most natural pollutants.

GROUNDWATER

Geology of Water in the Ground In general, precipitation that falls on the ground will (1) evaporate, or (2) run off into streams and lakes, or (3) trickle downward (infiltrate) into the ground. This last process, known as *recharge*, supplies groundwater to underground reservoirs. Groundwater can best be visualized as water filling holes or pores between mineral grains making up loose sand or solid rock. Obviously, if the grains are very smooth and spherical, and the spaces between them unfilled, there will be lots of room for water, and the *porosity* of the material is said to be high. If, however, the space between grains is filled with mud or some other cement, the porosity is low. If the water-filled holes are well connected so that water can flow readily through the material, it is said to be *permeable*. Any rock that has good porosity and permeability has the qualifications to be a good *aquifer*, that is, it has the potential to yield groundwater in large amounts.

Groundwater production is usually obtained from aquifers by drilling wells into them and pumping out the water. Almost all such wells pass through a zone of intermittently dry rocks before entering water-saturated rock at the water table. Because groundwater is derived from rainfall that seeps into the earth, the water table tends to reflect local topography, although it intersects the surface in springs and swamps. If wells are drilled deeper, the porosity and permeability of an aquifer are found to decrease progressively because of pressure from the overlying rocks. Thus, there is a relatively thin zone of groundwater in most parts of the world.

Problems Accompanying Misuse of Groundwater   When a well is drilled into a groundwater reservoir and pumping of water from wells is begun, care must be taken to balance the amount removed by pumping with the amount replaced by *recharge*. (Most recharge is natural, but it can be increased by pumping unneeded water down into other wells.) If such a balance is not maintained, problems such as ground subsidence, salt water encroachment, and eventually exhaustion of the water resource will be experienced.

Ground subsidence is caused most commonly by groundwater withdrawal, but can also result from removal of petroleum from shallow reservoirs such as those underlying Long Beach, California. Although arid and semiarid regions such as the Great Valley of California and the Valley of Mexico (City) provide impressive examples of subsidence due to groundwater withdrawal (Fig. 3-2), it can also be observed in humid climates if pumping rates are high. Houston and New Orleans, both of which receive abundant rainfall, have experienced subsidence of 2 to 4 feet within the last 25 years. In low-lying areas such as these, the threat of flooding increases rapidly with such subsidence. Recent estimates in the Houston area indicate that unless subsidence there is stopped, more than $1 billion worth of homes, as well as the Manned Spacecraft Center, will be submerged within a few decades. Drainage problems are also encountered in sewer and water systems. Furthermore, once the aquifer has been compacted by subsidence, it will not fully expand again. Thus, porosity and permeability are reduced and the aquifer can no longer serve as a good source of groundwater.

In many coastal aquifers, salt water, which is heavier than freshwater, extends landward beneath the freshwater. If freshwater pumping rates exceed local recharge, the salt water will move inland. This process, which is referred to as *salt water encroachment*, causes contamination of freshwater aquifers with salt water. The problem is most acute in cities such as Miami and New Orleans where good aquifers extend oceanward and pumping rates are high.

Exhaustion of fresh groundwater is the ultimate result of overuse, and the progress we have made toward this end is truly impressive. The most obvious examples of exhaustion of groundwater supplies are in arid regions. Here, recharge is small and groundwater usage is high because of the inadequacy of surface water supplies. Prior to settlement of the Llano Estacado region of western Texas, which is a rich agricultural area, the underlying groundwater reservoir contained about 60 trillion gallons of water that had accumulated over thousands of years. The present rate of water removal exceeds the average recharge by more than five times, however. This

Figure 3-2 Land subsidence due to overdraft of confined grounwater, Delta-Mendota Canal, western San Joaquin Valley, California. (*A*) Normal pipe crossing with pipe about 3 feet above water level. (*B*) Similar pipe crossing where land subsidence has lowered the pipe into the water. (*C*) More pronounced subsidence results in submergence of pipe. *(Photographs courtesy of N. Prokopovich and E. R. Lewandowski, U.S. Dept. of Interior, Sacramento, California.)*

has caused the water table to lower rapidly with a consequent increase in the difficulty of drilling for water and pumping it to the surface. Under these conditions, the West Texas groundwater reservoir is a nonrenewable resource (Chap. 1).

SURFACE WATER

Lakes and Rivers Surface water is the most important source of water to man not only because of its convenience but also because in many areas, such as the Piedmont of the southeastern United States, porosity and permeability in the underlying rocks are too low to permit production of adequate groundwater. In these areas, rivers and lakes are the principal sources of water. A fact that is not commonly appreciated is that lakes are freaks of the landscape. They form only under special conditions. Some lakes form where lava flows (Lake Yojoa, Honduras) and landslides (Hebgen Lake, Wyoming) block valleys, where volcanoes collapse (Crater Lake, Oregon), or where coastal processes form freshwater lakes (Lake Ponchartrain, Louisiana).

The only agents that form widespread and abundant lakes are glaciers and man. In North America, glacial lakes, including the Great Lakes, hold more than 30 percent of the total freshwater on the earth. In areas where natural lakes are not present, such as the southeastern United States, man forms lakes by blocking valleys with dams.

Problems Accompanying Misuse of Surface Water Because lakes are local depressions in the drainage system, rivers gradually fill them with silt and sand. Thus, many lakes (particularly artificial lakes) have useful life-spans of only a few hundred years, especially in arid areas such as the American West where rivers carry large loads of sediment. In addition to filling with silt and sand, lakes tend to change in chemical composition, which limits their usefulness as water sources. The Great Lakes, which attained their present form only a few thousand years ago at the end of the last glacial advance, provide an example of chemical degradation of lakes. Whereas the concentrations of sulfur, chlorine, sodium, and other elements heavily used by man have remained essentially constant during the last hundred years in more remote Lake Superior, the levels of these elements in the lower lakes, especially Lakes Erie and Ontario, have increased by as much as three times.

Overcommitment of surface-water resources and overuse of groundwater supplies usually go hand in hand. For instance, Baja California and Sonora in Mexico, and Arizona, Nevada, and Southern California in the United States, all of which are rapidly depleting their groundwater supplies, also vie for use of water from the lower Colorado River. Dams along this river form an almost continuous string of lakes along its lower 400 miles. In spite of carefully formulated agreements on how to share this water, the average annual flow of the river is only about 85 percent of the amounts committed to various areas.

LONG-DISTANCE WATER TRANSFER

As indicated in previous sections of this chapter, large parts of the United States and Mexico appear to be approaching rapidly a situation in which they require more water than can be obtained locally. In the year 2000, the water deficit for the United States alone is predicted to be about 85 billion gallons per day, or about one-quarter of the present daily water usage in the United

States. This water must be supplied from somewhere and, as in the case of other mineral resources, many people to the south look toward Canada and Alaska. Such a reaction is encouraged by the fact that there will almost definitely be a water surplus in western Canada and much of Alaska for the next few decades.

It is these observations that stimulated the proposal for a North American Water and Power Alliance (NAWAPA)—one of the most ambitious engineering schemes ever conceived. This proposal, which remains a gleam in the eyes of contractors and North American desert dwellers, called for the accumulation of large volumes of water in man-made lakes along the Rocky Mountain trench and other major valleys in western Canada and Alaska. These reservoirs were expected to yield about 120 billion gallons of water per day with 69 percent going to the United States, 13 percent to Mexico and 18 percent remaining in Canada for use in the Prairie Provinces. Water transfer was to have taken place via a complex system consisting of 6,700 miles of canals and 1,800 miles of tunnels.

Critics of the NAWAPA plan have pointed out that even this massive effort can do no more than cope with our expected water demand in the year 2000. Some Mexicans, quite justifiably, feel that if such a large North American effort is to be made, they should get more than the dying trickle from the end of the pipe. Many Canadians see in the plan yet another attempt to usurp Canadian sovereignty over their own natural resources. What would happen, for instance, if Canadian water demands exceeded predicted levels and some of the "exportable" NAWAPA water was needed for irrigation in the prairies? Could Canada divert these waters or would they be North American property?

As far fetched as the NAWAPA scheme sounds, it is not impossible and should not be ignored. Other mineral resources are transferred long distances. So, why not move water, if the consumer is willing to pay for it? The argument that people should move to water-rich regions is fallacious. Many water-poor but petroleum-rich areas such as Southern California and West Texas could level reverse arguments at residents of wet, but petroleum-poor areas. Precedents for long-distance water transfer exist. Southern Californians presently derive most of their water from a continuously growing system of aqueducts and tunnels stretching over 1,000 miles and tapping the Colorado River as well as streams on both sides of the Sierra Nevada mountains in central and northern California. International water agreements have been reached by Mexico and the United States in the Colorado River area and by the United States and Canada along the Columbia River. Thus, unless our rampant growth can be limited, it seems likely that larger and larger water-transfer projects will eventually make it possible to flush a toilet in Hermosillo, Mexico with last year's melt water from Mt. McKinley, Alaska.

FOUR

OUR COMMON MORTGAGE—
SOME CONSTRUCTION MATERIALS

INTRODUCTION
The construction industry consumes a tremendous quantity and variety of mineral resources. Unlike iron, aluminum, or copper, however, many of these mineral commodities are used in such a camouflaged form that their true importance is not readily apparent. This is unfortunate because construction is one of the few industries that affects almost all of us directly by governing the quality and cost of our shelter and, in many cases, by providing employment. Thus, we should be fully aware of the source of our construction raw materials, if we are to adequately house our growing North American population. Although we will discuss most important construction materials in this chapter, the geology of some of the materials is described further in Chapter 5.

STONE

Aggregate Aggregate is the industry term for fragmental rock material. Natural rock aggregate (Table 4-1) is available from deposits of sand and gravel or can be quarried and crushed to make crushed stone. Sand and gravel are usually cheaper because only washing and sizing are required to prepare them for market. They are found in stream and beach deposits (Fig. 4-1) and in deposits laid down by glaciers. Many areas of North America lack abundant sand and gravel deposits or have exhausted them. In some coastal areas, dredging of offshore materials (Chap. 10) supplements local supplies but the extremely low value of sand and gravel (Table 4-1) limits such offshore contributions to less than 3 percent of total U.S. consumption. Thus, it is commonly necessary to quarry and crush suitable rocks. Limestone (Chap. 5) makes up over three-quarters of all crushed rock because it is durable enough for most uses and yet soft enough to be crushed easily. Harder igneous rocks are used largely in areas where suitable limestone is unavailable.

Processed materials supply 2 or 3 percent of the aggregate consumed in the United States (Table 4-1). Slag, the largest source, is a waste product recovered largely from steel mill blast furnaces. Ash, the residue from combustion of finely ground coal, is recovered largely from power plants. Over 30 million tons of very fine fly ash are recovered annually from smoke stacks in North America and another 20 million tons of bottom ash are recovered from furnaces and boilers. About 20 percent of this ash is actually used now, mostly in road paving compounds.

Lightweight aggregate weighs about half as much as gravel or crushed stone. In general, the decrease in strength of lightweight aggregate is more

Figure 4-1 Sand and gravel operations, such as this one (S, P) in Arroyo Trabuco near San Juan Capistrano, are often located in stream valleys. Most such operations are begun when the surrounding area is rural (PS) but are gradually surrounded by new housing developments (A) *(From California Division of Mines and Geology.)*

Table 4-1 Sources, consumption, and approximate value of stone used in the United States in 1973. Over 85 percent of this stone was used in construction. The major source of nonconstruction stone is crushed limestone.

MATERIAL	1973 CONSUMPTION (millions of tons)	AVERAGE PRICE (per ton)	PRICE RANGE (per ton)
Aggregate			
Crushed stone (75 percent used in construction)	922	$ 1.60	$1.00 to $10.00
Sand and gravel (95 percent used in construction)	947	$ 0.60	$0.50 to $1.10
Slag	23		
Lightweight aggregate (95 percent used in construction)	18	$ 6.00	$3.00 to $60.00
Ash	10		
Dimension stone	2	$50.00	$20 and up

SOURCE: Cooper (1970*a* and *b*), Laurence (1973), Meikle (1973), and Nester (1973).

than offset by the savings from handling lighter material. Nature's lightweight aggregate is pumice, a volcanic glass in which the expansion of water vapor created innumerable tiny holes. Some pumice will actually float. Manufactured lightweight aggregate can be made from other glassy volcanic rocks, such as perlite and obsidian (which makes good arrowheads), some shale and clay, and even slag (plus water), by heating them to 1400 to 2000°F.

Dimension Stone Dimension stone (Table 4-1) is natural stone that is used in slabs or blocks for ornamental, building, and paving purposes. With the development of the rock crusher and portland cement in the mid-1800s, overall stone use gradually shifted from dimension stone to aggregate because of the lower cost of brick and mortar construction. Nevertheless, dimension stone offers aesthetic qualities that have kept it in demand. Most dimension stone is either limestone or granite with marble (metamorphosed limestone), sandstone, and slate (metamorphosed shale) bringing up the rear. With the exception of slate, which is valued for the flat slabs into which it splits, rock suitable for dimension stone production must occur naturally without closely spaced fractures or planes of weakness. To allow removal of large blocks, quarrying is usually done by channelling machines that cut grooves by impact, by endless wire saws that also cut grooves with sand as a cutting medium, and by expansion-wedging of closely spaced drill holes. Blasting, of course, cannot be used.

Although most dimension stone demand is satisfied by domestic sources, some stone, such as the well-known, iridescent black larvikite of Norway, is exported to many countries.

CEMENT AND CONCRETE

Cement (Limestone) Modern cement, known as portland cement, is made by heating finely ground limestone (80 percent) and shale or clay (20 percent) in a kiln. Cement clinkers, which emerge from the kiln, are mixed with about 4 to 5 percent gypsum (Chap. 5), to control the hardening rate, and pulverized to form cement powder. When mixed with water, cement hardens to form an artificial rock that withstands weathering as well as most natural rocks. Annual cement production reaches about three-quarters of a ton per person in some western European countries and averages about 200 pounds worldwide. Cement is manufactured in over 110 countries and though usually transported short distances, it can be shipped economically for several thousand miles in times of short supply.

Concrete (Aggregate and Asbestos) Concrete is artificial rock consisting of a mixture of aggregate and cement. Because cement costs about $20 a ton, it can be used more economically by mixing it with aggregate such as sand and gravel or crushed stone. Where weight must be minimized, lightweight aggregate is used.

Special fire- and weather-resistant concrete can be made by using about 15 to 20 percent asbestos as aggregate. About 70 percent of world asbestos production goes into cement products. Fireproof and lightweight plaster and vinyl tile account for most of the remainder of asbestos demand. By far the most widely used of the fibrous minerals collectively called asbestos is chrysolite (Appendix I), which forms strong, flexible fibers up to several inches in length. Long-fiber asbestos, which can be woven like wool, is rare and sells for $400 to $1,800 per ton. With the development of methods to bond shorter fibers (which cost only $35 to $400 per ton) economic interest has shifted to these lengths. Such short-fiber asbestos, when inhaled, can be carcinogenic. Because almost all asbestos reaches the consumer in bonded form, however, this danger is of concern chiefly in the asbestos mining and processing industries, where steps have been taken to closely monitor the air quality in the plants.

Asbestos is formed when circulating water reacts with magnesium-rich igneous rocks similar to those that are weathered to form nickel laterite deposits (Chap. 8). The resulting altered rock, known as serpentinite, commonly contains veins of asbestos fibers. Minable asbestos deposits measure at least 10 to 50 million tons of ore containing about 5 percent fiber content. The major sources of asbestos in the world are Canada and the USSR, which together supply almost 80 percent of world consumption. Whereas about 95 percent of Canadian production is exported, only 15 percent of USSR production is sold abroad, making Canada the main source of world asbestos supplies. Over 80 percent of Canadian production comes from a belt of serpentinite that follows the Appalachian Mountains through Quebec (Fig. 4-2).

PLASTER (GYPSUM)
Most plaster is made by heating gypsum (Chap. 5) to 250 to 400°F and driving off about three-quarters of its contained water. The resulting plaster will recrystallize as gypsum when mixed with water. Plaster is used to cover walls

Figure 4-2 Location of some important sources of the mineral commodities discussed in this chapter. With the exception of rutile, which comes largely from Australia, North America is well endowed with most of these mineral resources. In fact, deposits of structural clay, silica sand, dimension stone, crushed stone, and sand and gravel are too widespread to be shown on this map.

either as a paste or in the form of wallboard in which plaster is sandwiched between two sheets of heavy paper. Most plaster is combined with sand, lightweight aggregate, or short-fiber asbestos in both paste or wallboard. North American plaster production requires 15 to 20 million tons of gypsum annually, over two-thirds of total consumption, with the remainder being used in cement.

STRUCTURAL CLAY

Structural clays are used to make bricks, tile, sewer pipe, and other common construction products. Over 30 million tons of structural clay are consumed annually in these uses in North America and slightly less is used in cement and lightweight aggregate manufacture. Clays used for structural purposes must be readily heated, or fired, to form durable products. Color and minor impurities such as sand are not critical, and such raw materials are relatively widespread, and cheap ($1 to $2 a ton). In addition to structural clay there are many other types of clay raw materials. Ball clay and fire clay, for instance, contain few impurities and are used for special purposes such as brick for metallurgical furnaces, high-quality tile, and some whiteware. These clays are more expensive, ranging from $5 to $15 a ton, and annual production is about 10 million tons in North America. Fuller's earth and kaolin are clays with different uses as discussed later.

All clays are hydrous aluminum silicates with varying amounts of impurity elements such as sodium, calcium, and magnesium. Clay minerals form at and near the earth's surface either by weathering (Chap. 7) or by leaching processes carried out in shallow hot spring and hydrothermal systems (Chap. 8) such as at Yellowstone Park in Wyoming. Most structural, ball, and fire clays have been formed by weathering and accumulated in minable deposits in ponds, lagoons, broad stream valleys, or in shallow parts of the ocean.

GLASS

With the development of stronger glass and more innovative architectural styles, glass has displaced many other construction materials. Presently, about one-third of North American glass production goes into construction uses. Glass consists largely of the elements silicon and oxygen, which are usually obtained from quartz (Appendix I). Quartz is a common mineral in many rocks and is resistant to weathering. During weathering and erosion, it is usually concentrated in sand deposits formed by stream or surf action.

Quartz can be converted to glass by being melted and then cooled quickly before it can crystallize. To lower the melting point of quartz and to produce glass with special properties, a wide variety of elements and compounds are added to the quartz (Table 4-2). Among the most important is sodium, which is added in several forms. The most common form is soda ash, which is now prepared largely from the sodium-rich mineral, trona, which is found in enormous evaporite deposits in the Green River oil shale (Fig. 4-2 and Chap. 6). Other sodium is added as the mineral feldspar (Appendix I), which is recovered from igneous and sedimentary rocks and sand deposits, and from nepheline syenite, an unusual intrusive, igneous rock very rich in feldspar and low in iron (Fig. 4-2).

Lithium, which increases the strength and lowers the viscosity of glass, comes from both pegmatites and saline lake evaporite deposits (Fig. 4-2). Pegmatites, which are igneous rocks with very large crystals, are mined for lithium minerals (Appendix I), as well as feldspar and mica. Lithium-rich pegmatites in North Carolina, Rhodesia, and Canada are adequate to supply many decades of world lithium demand. The other source, saline lake evaporite deposits (Fig. 4-3), were formed where large inland lakes were partly dried up, leaving thick accumulations of evaporite minerals (Chap. 5) plus saline surface and groundwaters known as brines. Lithium, magnesium, sodium, potassium, and other elements are recovered from the brines at Searles Lake, California, Silver Peak, Nevada, and Great Salt Lake, Utah. Boron, which is used to make glass more resistant to heat, is mined from borate-rich (Appendix I) evaporite deposits around Death Valley, California.

PIGMENTS

Pigments supply color to paints, plastics, and other materials commonly used in construction. The iron oxide compounds ocher, umber, and sienna are natural mineral pigments commonly formed by weathering of iron-rich minerals, but most pigments are compounds manufactured from petroleum or the metals. Among the most important manufactured pigments are zinc compounds, which account for about 15 percent of North American zinc consumption, and carbon black, which is elemental carbon made from natural gas or petroleum liquids.

Table 4-2 Approximate amounts of material consumed in glass manufacturing in the United States in 1972. Some of this material is processed into other forms before being used.

MATERIAL	AMOUNT (thousands of tons)
Silica sand	11,000
Trona	3,000
Limestone	1,000
Feldspar	450
Nepheline syenite	350
Boron chemicals (including borax)	300
Barite	60
Lithium (as spodumene, a lithium silicate mineral)	10

SOURCE: Estimated from data of the U.S. Bureau of Mines; U.S. Geological Survey; and *Mining Engineering* (1973).

38 OUR FINITE MINERAL RESOURCES

Figure 4-3 Clayton Valley, near Silver Peak, Nevada. This large valley was a lake during the last major, worldwide glaciation. As the climate became drier, during the last 10,000 years or so, the lake evaporated leaving an accumulation of salts (white areas) and concentrated brines. This particular evaporite deposit is mined for lithium, which is recovered from the interstitial brine pumped from wells drilled into the white area. *(Courtesy of Thomas L. Kesler.)*

Titanium dioxide is an important white pigment derived from the minerals rutile and ilmenite (Appendix I). From one to three pounds of it are used in a gallon of light-colored paint, and more than 95 percent of annual world consumption of over 2.5 million tons is used for pigment production. Much of the remainder is used to make titanium metal, which competes with iron and aluminum for some aerospace applications. Ilmenite concentrates sell for about $20 to $40 a ton, and rutile concentrates, which contain twice as much titanium, command $150 to $250 a ton. Both rutile and ilmenite can be concentrated during the crystallization of igneous rocks rich in calcium feldspar. Because they resist weathering, they have also been concentrated in stream and beach placer deposits. Although ilmenite deposits are relatively abundant in North America (Fig. 4-2), we have no large rutile reserves and must import our supplies largely from Australia.

FILLERS
Fillers are fine-grained natural and synthetic compounds added to a product to substitute for more expensive material or to provide some special property. Aggregate and asbestos serve as fillers in concrete, for instance. Kaolin (Appendix I) is one of the most desirable fillers. Over 70 percent of

the 5 million tons produced annually in the United States is used to fill voids in raw paper sheets and to coat paper with a smooth, opaque white surface. This important use of kaolin derives from the fact that it, as well as most other clay minerals, occurs in the form of submicroscopic hexagonal plates, which can be spread to form smooth surfaces. Most of North American kaolin production comes from sedimentary deposits in central Georgia (Fig. 4-2), where it sells for $10 to $60 a ton. Fuller's earth, another clay product, was once used by "fullers" to clean and fluff wool and now finds applications in absorbent cleaning and sweeping compounds and in pesticides where the pesticide liquid is carried by fuller's earth powder. Fuller's earth is mined largely in Georgia and Florida (Fig. 4-2) and sells for about $20 to $30 a ton.

Talc (Appendix I) is another filler that finds minor use as a carrier in scented talcum powders and insecticides and major use as a filler in ceramics, paint, roofing, and rubber products. Because talc sells for $10 to $80 a ton, it is commonly used as an extender to minimize requirements for more expensive pigments such as titanium dioxide. The soft, slippery talc flakes are hydrous magnesium silicate that formed where dolomite, the magnesium-rich form of limestone, has been altered by silicate-bearing hydrothermal solutions (Chap. 8).

Mica, another filler, is familiar to us as the silicate mineral that forms small, shiny plates. The mica of chief commercial interest is muscovite (Appendix I), which is a common constituent of many igneous rocks, including pegmatites, and is usually mined from such rocks where weathering has destroyed the feldspar and freed the mica. About 150,000 tons of finely ground mica, which sells for $20 to $200 a ton, are used annually in North America as an extender in paints and as a filler in plaster, roofing products, and rubber. Natural and artificial calcium carbonate of very fine particle size is also an important extender.

ABRASIVES

This final aspect of construction minerals deals with the materials of relatively great hardness that are used to smooth, shape, and clean wood, plastic, glass, metal, and stone. For work on wood and some glass and stone, silica sand is adequate and over 600,000 tons of it are used annually for this purpose in North America. Other, harder, natural mineral abrasives include garnet (a complex silicate mineral), corundum (an aluminum oxide), and emery (a natural mixture of corundum and magnetite for which the emery wheel is named). These minerals are used in sand blasting, glass polishing, and nonskid pavements. Synthetic compounds, of which aluminum oxide is most important, have largely displaced these natural abrasives because they can be produced with more uniform size and hardness.

The only natural abrasive to withstand the competition of synthetics is diamond (Appendix I), which is the hardest material known and therefore is in heavy demand for a wide variety of industrial applications. Industrial diamonds include those diamonds that are too small or too low in quality to be gems. About 35 million carats (over 7 tons) of industrial diamonds come annually from the diamond-mining operations discussed in Chapter 9. A slightly larger amount is synthesized, largely in the United States, the USSR, Ireland, and Japan, by subjecting carbon to temperatures of about 5500°F and pressures over 70,000 times higher than those at the earth's surface. However, it has been possible to synthesize only very small diamonds so far, and demand for the larger, more expensive industrial stones is still supplied entirely by natural material. Presently known resources of natural industrial diamonds are probably not adequate to meet expected demand beyond 1985.

AVAILABILITY OF MINERAL RESOURCES FOR CONSTRUCTION

From the standpoint of geological distribution there is no reason to be concerned about the availability of the construction mineral resources such as aggregate and brick clay. In fact, this widespread availability and the resulting strong competition for markets means that the price of a ton of these materials is so low that they can be transported economically for only a short distance (Table 4-3). Hence, quarries must be developed as near as possible to highly populated areas where construction demand is high. This, of course, leads to conflicts between urban sprawlers and hapless quarrymen who supply the material for further urban sprawl. The quarryman's efforts to open new operations in the less-crowded countryside often result in confrontations with groups wishing to prevent areas of local scenic beauty from being pulverized.

The solution, though obvious, is not an easy one. To some extent, quarrymen have to stop setting themselves up on local promontories or bluffs which, though offering the easiest mining also have the greatest scenic appeal. Similarly, both citizens and government must allow cooperatively chosen space for quarries to operate economically at costs that urban growth can afford. The only alternative is found where water transport allows material to be barged to market.

At the other end of the scale are commodities such as ilmenite, mica, kaolin, asbestos, and lithium—minerals which are found in minable deposits in only a few areas. The prices of these commodities reflect their

Table 4-3 Range of freight rates (in 1970) for hundred ton lots of ore and concentrate on several common transportation systems. Actual rate depends in part on distance as well as tonnage. Note that water and pipeline transport are usually less expensive than rail or truck freight.

TRANSPORTATION SYSTEM	COST PER MILE (per hundred tons)
Ocean shipping	$0.03–$1.00
Pipeline	$0.15–$1.00
River barge	$0.20–$0.40
Rail	$0.40–$1.50
Truck	$5.50–$7.00

SOURCE: Maddex and Skaarup (1970).

Figure 4-4 The Jeffrey asbestos mine in eastern Quebec, the world's largest asbestos mine, removes 120,000 tons of ore and waste each day. As it has grown, it has impinged on the town of Asbestos (right) which has gradually been relocated. The waste and tailings from the mining and milling of the ore extend to the left-hand side of the picture. *(From Canadian Johns-Manville Co. Ltd., Asbestos, Quebec.)*

scarcity and permit their worldwide transportation and trade. Nevertheless, the mining of these commodities is not without problems locally. For instance, when the Jeffrey asbestos mine in Quebec (Fig. 4-4) was started in the 1880s, no one imagined that it might some day account for 13 percent of annual world asbestos production. Thus, it was logical that the town of Asbestos, where most of the mine's employees lived, should have developed as near as possible to the mine. As production approached today's levels of 600,000 tons of asbestos fiber per year, however, mining and milling operations gradually encroached on the growing town of 10,000 people, setting the stage for what could have been a confrontation. Instead, cooperative redevelopment of Asbestos, undertaken by the townspeople and the asbestos company, has resulted in a remodeled city in which very few houses are more than 15 years old and a mine that retains its rank as the world's largest asbestos producer.

FIVE

LEGACY OF PAST OCEANS—
OUR AGRICULTURAL AND CHEMICAL SUPPLIES

INTRODUCTION

Although today's oceans cover an impressive three-quarters of the earth's surface, geologic evidence suggests that ancient oceans were even more extensive. Most of these old oceans were fairly shallow, much like the present Hudson Bay, which averages only 420 feet deep. They advanced over the continents when the earth's crust was warped slightly or when the polar ice caps melted. While they flooded the continents, these shallow seas deposited many mineral resources essential to our agricultural and chemical industries (Fig. 5-1). The importance of the mineral resources discussed in this chapter and in Chapter 4 is clearly indicated by the fact that they are commonly known as industrial minerals. In highly developed countries, such as the United States, the value of industrial mineral production commonly exceeds that of metal ore production.

EVAPORITE DEPOSITS

As you know, sea water that evaporates on your skin leaves a scum of chemicals including a large amount of common salt. Experimental evaporation of sea water indicates that these chemicals appear in a specific

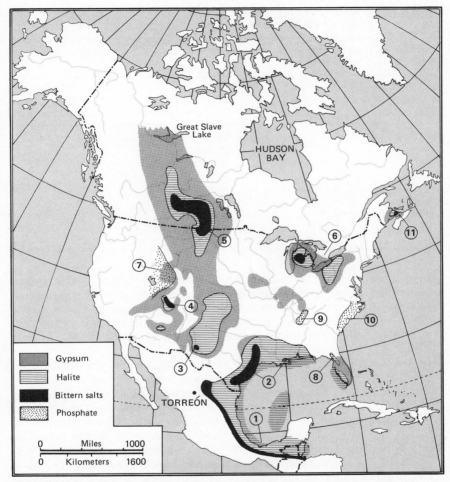

Figure 5-1 Distribution of evaporite and phosphate sediments in North America. The numbers indicate the location of the following areas mentioned in this chapter: (1) southern Mexico salt dome-sulfur area; (2) Gulf Coast salt dome-sulfur area; (3) southeastern New Mexico (Permian) potash area; (4) southeastern Utah (Paradox) potash area; (5) Saskatchewan (western Canada) potash area; (6) Michigan-New York-Ontario salt area; (7) western U.S. (Phosphoria) phosphate area; (8) Florida phosphate area; (9) Tennessee phosphate area; (10) N.C.-S.C. phosphate area; (11) Maritimes gypsum, salt, potash area.

sequence and in predictable amounts (Table 5-1). Not too surprisingly, a similar depositional sequence is followed when entire shallow seas evaporate. The chemical deposits formed in such seas are appropriately called evaporites and they are very abundant in North America (Fig. 5-1). Many evaporite deposits are hundreds of feet thick and extend over hundreds of

Table 5-1 Sequence of deposition and relative abundances of minerals deposited by evaporating sea water. Most gypsum is actually deposited as the mineral anhydrite ($CaSO_4$) which combines with water to form gypsum at lower temperatures. The most important of the bittern salts is sylvite (KCl), which is the most common source of potash.

SEQUENCE OF DEPOSITION	MINERAL(S)	COMPOSITION	PERCENT OF TOTAL SOLIDS DEPOSITED
First	Calcite	Calcium carbonate, ($CaCO_3$)	0.3
Second	Gypsum	Hydrous calcium sulfate ($CaSO_4 \cdot 2H_2O$)	4.5
Third	Halite	Sodium chloride (NaCl)	77.2
Fourth	Bittern salts	Magnesium, potassium, and bromine compounds	17.9

square miles, which is far too much material to have formed by evaporation of a sea only a few hundred feet deep. This problem, which has perplexed geologists for years, must have its answer in some process by which a shallow, almost isolated bay undergoes rapid evaporation and subsidence while being replenished with new sea water through a small inlet.

Gypsum and Sulfur Gypsum is a hydrous calcium sulfate that is used largely in plaster and portland cement, two very important construction materials (Chap. 4). During the last few years, world gypsum production has amounted to about 60 million tons, of which North America has produced and consumed about half. As Fig. 5-1 indicates, gypsum-bearing evaporite deposits are very extensive in North America and there is little reason to worry about future supplies. In spite of this abundance, however, over 35 percent of U.S. gypsum consumption is imported. The reason for this is transportation costs, a big concern to producers of many inexpensive mineral commodities. Gypsum is worth about $4 to $6 per ton, a price so low that only a small distance of overland transport (Table 4-3) can cause a significant percentage increase in the price. This, of course, shifts the economic advantage to deposits that are close to the consumer or those that are located near water transportation. Thus, deposits in Nova Scotia and Jamaica can supply gypsum to the construction industry along the U.S. east coast more cheaply than can domestic deposits to the west.

Gypsum consists in part of sulfur (Appendix I), one of the mainstays of most industrial processes. (Most sulfur is used to make sulfuric acid, much of which is used to treat phosphate fertilizers, which are discussed later.) However, there is no widely applied inexpensive method of separating sulfur from gypsum. Fortunately, nature has done this for us locally, and much of our sulfur production comes from gypsum accumulations that have undergone this natural process. (Table 5-2).

Table 5-2 Sources of sulfur production (as a percentage of total production) in the United States, Canada, and Mexico. All forms of sulfur (native, sulfuric acid, hydrogen sulfide, sulfur dioxide) are included.

SOURCE	PERCENTAGE OF TOTAL PRODUCTION		
	MEXICO	UNITED STATES	CANADA
Native sulfur from evaporite deposits	94	76	—
Natural gas with hydrogen sulfide	4	7	81
Metal sulfide ores	—	8	19
Other sources (largely volcanic deposits)	2	9	—

SOURCE: U.S. Bureau of Mines.

Strange as it seems, there are some bacteria in oxygen-poor areas such as swamps or groundwater that will transform the sulfate ($SO_4^=$) ion (from dissolved gypsum) into hydrogen sulfide gas (H_2S). Some of this gas will escape to the atmosphere, but part of it will be oxidized to produce solid, elemental sulfur, called native sulfur. Many large sulfur deposits formed by this process in the Gulf Coast area and in West Texas are at depths of 500 to 3,000 feet and contain about $1 to $8 worth of sulfur per ton of rock. This grade-depth combination will not permit any type of conventional mining to be done economically. Fortunately, however, sulfur can be melted *in situ* by forcing superheated water and compressed air from the surface down into the ore zone and pumping the resulting froth of air, water, and melted sulfur back up again. This process is so efficient and simple, in fact, that it can be used on offshore platforms similar to those used in the oil industry (Chaps. 6 and 10).

North American reserves of native sulfur total at least 300 million tons, mostly in Texas, Louisiana, and eastern Mexico. At least that much is in Canadian natural gas (Chaps. 1 and 6) and sulfide mineral deposits (Chap. 8). At present annual world consumption rates in the range of 35 million tons, these supplies are adequate for at least two decades. Backing up these reserves are the largely uneconomic but huge amounts of sulfur that might be recovered from crude oil and coal, and, of course, all the gypsum in the world.

Salt Annual world production of salt amounts to almost 100 pounds per person, only a small fraction of which goes on your fried eggs. The rest finds a variety of uses ranging from the manufacture of hydrochloric acid, an essential industrial acid, to snow removal. Most of the world's salt comes from the mineral halite, which is usually mined from evaporite deposits by underground methods. Many mines, such as those in Ontario and New York (Fig. 5-1), operate on relatively flat layers of salt tens of feet thick and only a few hundred feet below the surface.

In the Gulf Coast region, however, thick salt layers are thought to be at depths of up to 40,000 feet, far too deep for economic mining. The evidence for this salt layer is in the form of salt domes (Fig. 5-2), peculiar vertical cylinders or blobs of salt that have penetrated great thicknesses of overlying sediment. Because the salt has a lower density than that of the surrounding sediment, it is actually floating upward through the sediment and can even extend above local land level. The five "islands" of southern Louisiana, including Avery Island, Jefferson Island, and Weeks Island, are salt domes with shallow mantles of soil. The tops of these salt domes have risen over a hundred feet above the local land surface, much of which is swamp at near sea level, hence the term "island." Underground salt mines are active in these domes, and much of the salt goes up the Mississippi River on barges that are used to carry midcontinent grain to New Orleans.

Groundwater circulating around some of these shallow salt domes has

Figure 5-2 Avery Island in southern Louisiana was formed when a salt dome pushed the surficial sediments as much as 150 feet above the surrounding swamps. The salt dome itself comes to within 16 feet of the surface and is thought to have come from a "mother salt" bed at a depth of 5 to 7 miles below the surface. About 10 million tons of salt have been removed from the dome by underground mining, and oil is found along the sides of the dome at depths between 4,500 and 15,600 feet. The fields on the left-hand side of the photo are the principal world source of chili peppers for Tabasco sauce, and a popular tourist attraction, Jungle Gardens, is located at the bottom of the photo. The Gulf of Mexico can be seen in the upper left-hand corner of the photo. *(From International Salt Co., Clarks Summit, Pennsylvania.)*

dissolved and removed salt from their upper parts and has left an accumulation of less soluble gypsum and anhydrite. Bacterial decomposition of a few of these sulfate-rich salt dome cappings has produced the Gulf Coast sulfur deposits mentioned earlier.

Unpurified rock salt mined from evaporite deposits costs about $4 to $8 a ton at the mine and is not transported long distances except under unusual circumstances such as mentioned above. Over 15 percent of the continental area of the earth is underlain by evaporite salt, much of which is minable, and so long-distance transport of salt is not usually required. Along with sea water and brines (Chap. 4) evaporites constitute an essentially inexhaustible supply of salt.

Potash Potash, which is the industry term for potassium chloride, is one of the three most important constituents of fertilizer. (The others are phosphates, which we will discuss later, and nitrate, which is a nitrogen-oxygen compound once mined in Chile but now derived largely from the atmosphere.) The most common present mineral raw material for potash is one of the bittern salts, sylvite, which is potassium chloride. The term *potash* remains from several centuries ago when potassium was obtained by leaching plant ashes with water in an iron pot. The fact that potassium was plentiful in plants eventually led to its recognition as an essential element in plant development.

The three major potash-mining areas in North America are southeastern Utah, southeastern New Mexico, and southern Saskatchewan (Fig. 5-1) of which the last two have been the most important producers. Although the total amount of potash, its grade, and the thickness of the ore layers are lower in the New Mexico deposits, the Saskatchewan deposits are at depths of 3,000 to 4,000 feet—about twice that of the New Mexico deposits. This extra depth is a real disadvantage. For one thing, several aquifers between the surface and the ore zone contain water under high pressure that will flood the mines if it is not sealed out. Furthermore, the weakness of the potash ore under pressures found at depths of 3,000 feet causes support pillars in the mine to flow readily if cut too thin. This, of course, can cause the roof (called "back" by miners) to collapse, forming fractures that will let in water to flood the mine. Thus, large support pillars must be left in the mine and, since they consist of ore, over half of the ore stays in the ground. In direct contrast, the New Mexico miners can recover almost all their ore by robbing pillars, a process by which the original support pillar is almost totally removed as the miners retreat from an area of the mine that is to be abandoned.

The massive mining problems of the Saskatchewan and Utah potash districts have encouraged the development of solution-mining techniques, a process whereby water is pumped down into the ore zone, where it dissolves the ore, and is pumped up for processing. This method, however, is largely undoing the job that nature did for us by evaporating the ocean in the first place. Thus, solution mining can require more expensive processing that offsets the savings from cheaper mining. It also drives governments mad trying to decide whether the operation should be taxed as a well or a mine!

Potash has sold for about $20 to $40 per ton (of sylvite) during the past few years and has had a price history similar to that outlined for sulfur in Chapter 1. Known U.S. evaporite potash reserves are about 900 million tons (expressed as K_2O), which, though large in relation to present annual world production of 25 million tons, is minor when compared to recoverable Canadian reserves of about 10 to 15 billion tons. Similar large evaporite reserves are present in the USSR and Europe and in saline lakes and lake deposits such as Searles Lake, California and the Dead Sea (Chap. 4). Don't lose any sleep about our potash supplies.

OTHER MINERAL RESOURCES DEPOSITED FROM SEA WATER
There are a lot of mineral resources that we can attribute to the shallow seas. Some petroleum and natural gas, for instance, and even many sand and gravel deposits had their origin here. These commodities find better homes in other chapters, however, leaving us concerned here with mineral resources that formed by chemical precipitation from sea water, such as limestone and phosphate. Both of these compounds are closely linked to sea life of the past since limestone makes up most shells, and phosphate makes up bones and teeth.

Limestone Limestone is made up of small grains of calcium or calcium-magnesium carbonate. If the proportion of magnesium carbonate approaches 45 percent, the rock is dolomite. At least 800 million tons of limestone and dolomite are produced annually in North America. (What did you do with your three tons?) About two-thirds was used as crushed stone for roads (called *road metal*) and in concrete (Chap. 4). The remainder went into the manufacture of cement, into flux for steel mills, and into agricultural, chemical, and refractory uses.

Most of midcontinent North America from Torreón to Great Slave Lake (Fig. 5-1) has large limestone layers at or near the surface. Many of these layers formed in shallow seas similar to those that now surround Florida and the Bahamas. Almost all of this limestone originally formed part of the shell or skeleton of some organism and was deposited in sedimentary layers after the organism died.

Limestone suitable for use as crushed stone is abundant, and such limestone quarries and their attendant environmental problems are widespread (Chap. 4). For use in the chemical and metallurgical industries, however, limestone or dolomite must contain only about 3 percent or less impurities and must usually be relatively pure limestone or dolomite. Such limestone and dolomite layers are not abundant. Nevertheless, there is no foreseeable shortage of these materials.

Phosphate Along with potash and nitrate, the other major fertilizer material is phosphate, which is the industry term for minerals rich in the element phosphorus. It occurs in most ore deposits as the mineral apatite, a complex calcium phosphate, formed during crystallization of some igneous rocks, but deposited in minable concentrations more commonly when it precipitates from the sea.

What caused these phosphate deposits to precipitate from the sea? Well, it is known that sea water at depths below 3,000 feet has a high content of dissolved phosphate, whereas water near the ocean's surface is very low in phosphate. It seems that the change in temperature and pressure associated with the rise of deep water to the surface causes it to lose its load of dissolved phosphate. About 250 million years ago, for instance, a shallow sea covering eastern Wyoming (Fig. 5-1) deepened rapidly toward the west. Upwelling of sea water along this submarine rise is thought to have caused deposition of phosphate in the Phosphoria formation, a sequence of layers of limestone, shale, sandstone, and phosphate in which several individual layers, as much as 9 feet thick, contain 8 to 16 percent phosphorus. Phosphates formed in a roughly similar way are found in primary bedded deposits in Florida and North Carolina, and in residual deposits in Tennessee (Figs. 5-1 and 5-3).

Phosphate rock, as it comes from these deposits, is not in a form that is easily enough dissolved to be used directly as a fertilizer. In order to transform it into a suitable form, it is commonly digested in hot sulfuric acid and processed to form products such as triple superphosphate (TSP), which contains about 20 percent phosphorus. Since it takes about 1 ton of sulfur (as acid) to produce about half a ton of phosphate fertilizer, the phosphate industry is an enormous consumer of sulfur.

Annual world phosphate consumption is about 80 to 110 million tons (of concentrate known in the industry as phosphate rock) valued at about $10 to $15 per ton. This level of consumption is considerably higher than consumption estimates made for today by informed sources in the late 1950s, and

Figure 5-3 Phosphate mining at the Lee Creek Mine, North Carolina. Ore is moved out of the pit with the two draglines, then made into a slurry by the three water jets (right center), and pumped to the mill for processing. Size of the draglines can be estimated from the people in the right foreground. *(From Texasgulf Inc., Toronto.)*

concern has been expressed about the rate at which we might be depleting our available supplies. Fortunately, this concern is misplaced. Present estimates of global reserves and resources of phosphate rock range from 100 to 1,000 billion tons. This includes over 10 billion tons of relatively easily minable material in North America.

MISUSE OF THE OCEAN'S LEGACY

Excluding limestone and gypsum, which are relatively harmless environmentally, we use almost 300 million tons of the ocean's legacy each year and dump much of it back onto the earth's surface. What effects is this having? Take salt, for instance. Over 10 million tons of it are dumped on roads and streets each year in northern North America. This is a large amount even in relation to the 50 to 500 million tons of natural salt (from airborne sea spray) dumped on the continents by rain each year, especially in view of the very localized urban areas in which de-icing salt is used. In the Great Lakes area where de-icing salt is used heavily there have been very easily observed increases in the salt content of surface water (Chap. 3). Perhaps more importantly, there is growing evidence from carefully controlled local studies of salt balance that much of the de-icing salt is being retained in the groundwater and soil. There seems little doubt that this salt will have long-term effects on areas in which it is heavily used. The only question is, will these effects be tolerable in relation to the obvious benefits of safer winter driving conditions?

Sulfur, largely in the form of SO_2, is a major air pollutant. Combustion of sulfur-bearing fuels and, to a lesser extent, smelting of metal sulfide ores are presently releasing almost half as much sulfur into the air as are all natural processes combined (Table 5-3). Because the amount of sulfur put into the atmosphere by man each year is almost twice that of world sulfur consumption, collection of the sulfur before it enters the atmosphere would cause a huge increase in world sulfur production and a depression in the present sulfur industry.

Fertilizers, along with high-yield crops, have brought about the "green revolution," an unprecedented increase in agricultural productivity. That this revolution has much farther to go is indicated by the fact that per capita use

Table 5-3 Sources and sinks (where the sulfur goes) of sulfur in the atmosphere expressed as millions of tons of sulfate (SO_4)

SOURCES		SINKS	
Biological processes	268	Rain over oceans	217
Wind-blown sea salt	130	Rain over land	258
Volcanoes	2	Plant uptake and dry deposition	75
Fossil fuel combustion	150		
TOTAL	550	TOTAL	550

SOURCE: Kellogg, Cadle, Allen, Lazarus, and Martell (1972, Fig. 2).

of potash fertilizer in South America and Africa is less than 10 percent of that in Europe and North America. Many scientists question whether fertilizers can continue their role as mainstays of the green revolution, however, in view of the increasing evidence of their accumulation in the environment. Phosphates and nitrates are the greatest offenders along these lines because they are intimate parts of the life cycle. When present in excess, they hasten eutrophication, a process by which organic activity in surface waters is overstimulated causing a lowering in dissolved oxygen in the water and a consequent disruption of normal life patterns.

There are other major sources of phosphates and nitrates, such as sewage and automobile exhaust, of course, but fertilizers contribute to the problem because they are applied in easily soluble forms and only a small percentage of each application is actually used by the plants. The remainder goes into surface and groundwater. In view of these considerations, the use of fertilizers may have to be curtailed locally in the future or combined with organic or inorganic agents that will limit their loss from the soil.

SIX

ENERGY RESOURCES—
THE RISE OF THE ELECTRIC TOOTHBRUSH

INTRODUCTION

We need an energy appreciation day. There can be no question that energy, along with the structural metals, forms the basis for our industrial civilization (Fig. 6-1). The question is how to show our appreciation. You could leave your car, television, and electric toothbrush running all day and create a colossal brownout or gasoline line. A more memorable way, however, would be to stop all energy-consuming machines in the world for a day and to have each person explode 80 pounds of TNT. Daily world energy consumption would remain the same; we would simply use it in a different way. In the United States, where about one-third of the world's energy is consumed, such a celebration would require over 400 pounds of TNT for each person.

At this point in human evolution, a very large part of the energy we use is derived from mineral resources. In the United States, for instance, over 90 percent comes from the fossil fuels—coal, oil, and natural gas. Another few percent comes from nuclear reactors and naturally heated geothermal water systems. Power can also be derived from falling water (hydroelectric), the tides, the wind, and solar radiation, none of which will be described here. Our task in this chapter is to better understand the mineral resources on which

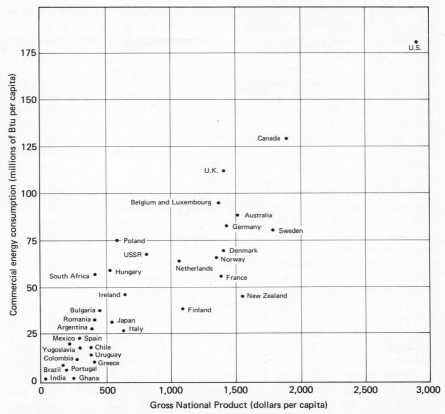

Figure 6-1 Relations between Gross National Product and commercial energy consumption in 34 countries of the world showing the close connection between energy and affluence. *(After Cook, "The Flow of Energy in an Industrial Society." Copyright © 1971 by Scientific American Inc. All rights reserved.)*

we are strongly dependent for energy and to see how much longer they can serve us.

FOSSIL FUELS

Plants and animals contain abundant hydrogen and carbon (hydrocarbons) and are combustible. Whereas most of the world's ancient life has died and decayed with hardly a trace, a small amount of it remains as fossil hydrocarbon accumulations in the form of coal, oil, and natural gas (Table 6-1). These are the most convenient forms, but there are many other mineral resources, such as oil shale, from which it is possible to make synthetic oil and gas. A very important feature of the fossil fuels is the fact that their presence depends on life. Thus, almost all the world's fossil fuels are found in sedimentary rocks deposited during the last 10 percent of the earth's history (Chap. 1). This eliminates the large areas of Precambrian rock (Fig. 7-2) from consideration, thereby rendering many South American and African nations poor in fossil fuels.

Coal Coal is a black, messy rock that will burn. This amazing property created a demand for about 650 million tons of coal, valued at nearly $5 billion, in North America during 1973. Most coal consists of the hardened remains of partly decayed vegetation known as peat. Beds of this material were deposited in ancient swamps where they were protected from the atmosphere, first by swamp water and later by layers of sediment that washed over the swamp. When peat is more deeply buried, it loses water and gases to form the carbon-rich rock known as coal.

Coal beds being mined today range from about 2 to 100 feet in thickness and average about 5 feet. About half of North American production comes from underground mines at depths less than 1,000 feet in which recovery averages 50 percent. Where the coal bed is not more than 50 to 100 feet below the surface, the overlying rock can be removed to recover almost 100 percent of the coal by strip mining. Because coal layers are extensive, tracts subjected to strip mining (Figs. 6-2 and 7-3) commonly cover many acres and attract more attention than most open pit mines. Present strip mining operations can restore the land surface almost to its original condition but do disrupt shallow aquifers. Contour mining is a type of strip mining done in hilly areas. It involves the extraction of as much as possible of a horizontal bed of coal that is exposed around the hillside. Contour mining allows economic recovery of coal layers too thin to be recovered by conventional strip mining.

Three factors determine coal quality. First is rank, which governs the energy released (Table 6-1). In general, higher rank bituminous and anthracite coals are in older rocks and, by a quirk of North American geology, most of these higher rank coals are in the East. Another factor controlling coal quality is ash, the incombustible sediment washed into the original peat and remaining as dust and cinders after burning. Ash is an air pollutant and a construction material (Chap. 4). Finally, the most important factor is sulfur, which can constitute up to 7 percent (by weight) of coal. When coal burns, this sulfur enters the atmosphere as sulfur dioxide, an atmospheric pollutant (Table 5-3).

Although eastern North American coals are the best fuels from the standpoint of energy content, they are rich in sulfur. Most of this sulfur is present as the iron sulfide, pyrite, which can be separated by beneficiation of the coal. Other forms of sulfur, such as sulfate compounds and organically bound sulfur, are less easily removed, however, and most Eastern coal retains about 2 percent sulfur after beneficiation. In contrast, lower-rank coals in the northern Great Plains and the Prairie provinces contain less than 1 percent sulfur before beneficiation. In the face of growing regulations limiting the sulfur emissions of coal-fired power plants, low-sulfur Western coal has attracted widespread economic attention. Even under favorable conditions, transportation charges add about 60 percent to the $5 to $8 per ton that most coal costs at Eastern mines. Thus, consumption

ENERGY RESOURCES—THE RISE OF THE ELECTRIC TOOTHBRUSH   55

Figure 6-2 Strip mining can be used to recover any tabular ore body that is near the surface and roughly horizontal, and is used very commonly in coal mining. In such an operation, the worthless overlying rock (overburden) is removed by a large machine such as a bucket-wheel excavator (A) or a dragline (B). The coal seam is then exposed on the floor of the excavation and is mined by a smaller shovel (lower rear in A, right in B) and hauled out by truck. Mining progresses along a large ditch or strip with waste piled on one side (right in A, left in B) and new rock exposed on the other side. Note the two dark, horizontal coal seams exposed in the wall of the pit in B. (From Mining Engineering.)

Table 6-1 Approximate heat content and carbon-hydrogen ratio of coal (free of ash contaminants), oil, and natural gas. One barrel (bbl) of crude oil equals 42 U.S. gallons and 34.97 Imperial gallons and weighs about 300 pounds. One cubic foot of natural gas (measured at a pressure of 14.73 pounds per square foot and a temperature of 60° F) equals 0.178 bbl.

	RANGE OF HEAT CONTENT (Btu per pound)	CARBON-HYDROGEN RATIO
Lignite coal	6,000–7,500	14
Subbituminous coal	8,500–11,000	14–20
Bituminous coal	12,000–15,500	
Anthracite coal	13,500–15,500	27–31
Crude oil	18,000–20,000	5–8
Natural gas	18–22	3

SOURCE: Tiratsoo (1972, Tables 2 and 12) and Averitt (Fig. 15 in U.S. Geological Survey Prof. Paper, 820, 1973).

of Western coal in the East would require cheap transportation such as piplines with a coal–water slurry or long "unit" trains. Alternatively, power can be generated at the coal mine for transmission to some consuming centers as is done in the Four Corners area in the southwestern United States.

Crude Oil and Natural Gas Crude oil is a complex mixture of hydrocarbon structures or molecules of varying size and complexity, all of which form liquids under normal conditions in the earth's crust. There are also some hydrocarbon molecules, dominantly methane (CH_4), which form natural gas. Both crude oil and natural gas are found largely in porous and permeable sedimentary rocks such as sandstones and buried limestone reefs where the hydrocarbon fluids fill the pores. Such rocks are known as reservoirs, and in them the oil and gas can be mixed, the gas can occur above the oil, or the two can be widely separated. Both gas and oil, however, are lighter than groundwater, and therefore they will tend to float upward and seek a surface outlet. Some such outlets are observed, such as the La Brea tar pits in California. In many cases, however, upward migrating oil and natural gas are trapped underground where porous reservoir rocks are sealed by impermeable rocks. The classic oil and gas trap is an arch or anticline of porous sandstone covered by impermeable shale. Other common traps form where good reservoir rocks are faulted against impermeable rocks or where they have been tilted up against the flanks of salt domes (Chap. 5).

Good oil and gas reservoirs must have high porosity and permeability. However, porosity decreases rapidly with depth in the crust as higher pressures close the openings in the rock. In fact, all the world's giant oil fields are at depths of less than 6,500 feet, and very little oil and gas production comes from depths of more than 25,000 feet.

Most of the sedimentary rocks from which oil and gas are produced were deposited in the ocean, not in freshwater lakes. The reservoir rocks from which production actually comes, however, do not appear to have contained much original organic matter. Instead, it is thought that the oil and gas were sweated out over long periods, from adjacent shales, which are less permeable but commonly contain a few percent of organic material. After it forms, crude oil gradually alters toward lighter crude compositions that are more valuable for their higher potential gasoline yield. A similar effect is produced by the higher temperatures accompanying deeper burial in the crust.

Oil and natural gas are removed from the ground by drilling holes into the reservoir rock and pumping. Because oil adheres to rock, an average of only about 30 to 35 percent of the oil in most reservoirs can be removed. Modern techniques involving careful control of the pumping rates, injection of gas or steam, and even burning part of the oil in the ground can improve this figure locally, but the fact remains that over half of the oil in many reservoirs never leaves the ground. It has been proposed that recovery could be increased by strip mining of oil fields, but this would be limited to the shallowest fields. Natural gas, being less viscous, can be recovered at a level of about 85 percent.

Refining of crude oil depends on the fact that different hydrocarbon molecules will boil off or distill at different temperatures. Some of these compounds, such as kerosene, are usable without further change, whereas others must be "cracked" by use of heat and pressure to produce lighter products such as gasoline. Some crude oils, notably those from Venezuela and Mexico, are high in sulfur, which must be removed. Natural gas processing involves separation of methane from other hydrocarbon gases such as ethane and from helium and the hydrogen sulfide that forms "sour" gas. Until recently, natural gas could be marketed economically only from reservoirs near urban markets, largely because its low energy content (Table 6-1) precluded transportation. Thus, large amounts of natural gas byproducts from crude-oil extraction and refining were disposed of by burning at the well or refinery. With the development of more efficient pipelines, as well as liquefied natural gas (LNG) tankers (in which gas can be shipped at −258°F in about 0.16 percent of its normal volume), gas can compete with other fuels on a worldwide basis.

During the past two decades enormous quantities of oil and gas have been discovered, and many previously ignored areas are now major producers. The Niger delta, originally explored because of its similarity to the prolific Mississippi delta, has made Nigeria an important source of low-sulfur oil to Europe. Offshore drilling is accelerating in many parts of the world and even the forbidding North Sea has yielded Norway one of the world's 50 largest oil fields. Costly and generally discouraging exploration in northern Alaska, which began in the early 1940s finally yielded the Prudhoe Bay discovery (Fig. 6-3) which has been estimated to be the largest single oil field in North America.

Notwithstanding the size of the Prudhoe Bay discovery and other developments now in progress, there seems little possibility that the present balance of world oil and gas reserves (Table 6-2) will change greatly. This is because the "giant" oil and gas fields contain most of the known reserves. Of the approximately 5,000 fields in the world, over 85 percent of production comes from only 5 percent of the fields and 15 percent comes from only two fields, Ghawar in Saudi Arabia and the Burgan complex in Kuwait. In 1970, 25 of the 40 largest fields were in Iraq, Iran, Saudi Arabia, Kuwait, and Bahrain.

As the 1973 to 1974 oil embargo indicated, such an extreme localization of mineral resources can become a deciding factor in international politics. In addition to the direct impact of supply embargoes, such resource monopolies can lead to massive distortions in international trade balances. In 1972, even before crude oil prices had tripled to today's levels of $10 to $15 a barrel, Saudi Arabia received over $3.1 billion in taxes and royalties for its oil and gas. This amount exceeds the GNP of many nations in the world and has made Saudi Arabia a major factor in the politically sensitive area of international finance. Although oil is perhaps more easily applied to such ends than other mineral resources, it is by no means unique. As supplies diminish and demand increases, we might expect similar situations to develop in aluminum, chromium, and other commodities.

The oil and gas industry has a long history of production problems going back to at least 1862, when about 5 million barrels of oil from a shallow well in

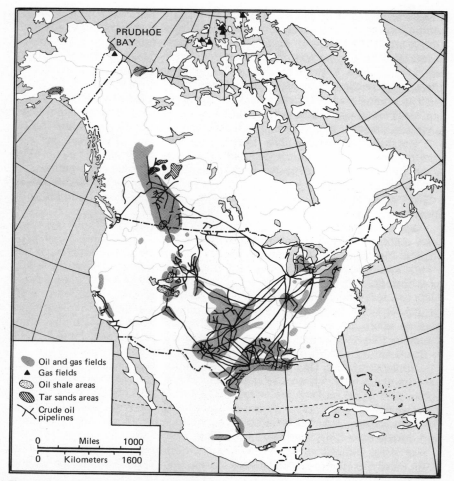

Figure 6-3 Distribution of areas containing crude oil, tar sand, and oil shale. Natural gas producing areas correspond in general, with known oil-producing areas except (so far) for the areas indicated in the Canadian Arctic islands. The generalized distribution of crude oil pipelines is also shown. Two planned pipelines, the Alaskan line and the Montreal line, are also shown.

southern Ontario covered Lake Erie. Gushers and similar problems at the wells became less common when heavy "muds" and special valves were incorporated in the drilling equipment.

Major attention today is focused on spillage during transportation, particularly marine transportation. About 60 percent of world oil production moves by sea each year. The abundant clots of heavy oil littering the surface of the oceans are commonly thought to be the heavy residue of oil spills and leaks. However, crude oil also enters the ocean from natural submarine

Table 6-2 Reserves of crude oil and natural gas in major producing and consuming countries (Jan. 1, 1974)

COUNTRY	OIL RESERVES (billions of bbls)	GAS RESERVES (trillions of ft³)
Saudi Arabia	132.0	50.9
USSR	80.0	20.0
Kuwait	64.0	32.5
Iran	60.0	270.0
USA	34.7	247.0
Iraq	31.5	22.0
Libya	25.5	27.0
Abu Dhabi	21.5	12.5
Nigeria	20.0	40.0
China	20.0	20.0
Neutral Zone	17.5	8.0
Venezuela	14.0	42.0
Indonesia	10.5	15.0
United Kingdom	10.0	50.0
Canada	9.4	50.3
Algeria	7.6	105.9
Syria	7.1	0.7
Qatar	6.5	8.0
Ecuador	5.7	5.0
Oman	5.2	2.0
Egypt	5.1	4.2
WORLD TOTAL	627.9	2,033.4

SOURCE: *Oil and Gas Journal*, 1973.

seepages in many parts of the world and the relative importance of the human and natural contributions has not been measured.

Further concern centers on pipeline transport, especially in the Arctic (Fig. 6-4) because oil spilled in cold climates will not break down rapidly. The major controversy concerns the fact that Arctic soils are permanently frozen and, when thawed, clay-rich soils can form a strengthless slurry. Such materials will not support a hot oil pipeline, particularly in earthquake zones such as characterize Alaska. Such problems, however, seem certain to be resolved by technological developments ranging from well-insulated pipelines with many safety valves to cold sea water–oil suspensions in the pipeline.

Synthetic Hydrocarbon Fuels There are many forms of natural hydrocarbons that can be processed to yield a synthetic oil or gas. Coal, for instance, can be heated and combined with various mixtures of oxygen, hydrogen, or water to produce synthetic gas similar to natural gas. This would be particularly attractive if it could be carried out underground to avoid strip mining. Attempts to do this are underway.

There are several rocks that can be treated to yield synthetic oil. Tar

Figure 6-4 Aerial view of the drill rig at Imperial Oil's well Atkinson H-25, 50 miles northeast of Tuktoyaktuk in the N.W.T. of Canada. This well was completed on Feb. 26, 1970 to a depth of 5,941 feet. Oil was found at about 5,700 feet in Cretaceous age (130 to 65 million years old) sands. *(From Imperial Oil, Ltd., Toronto.)*

sands, which are presently being exploited, are porous sandstones filled with a solid hydrocarbon residue called bitumen. It is generally accepted that this bitumen represents the residue remaining after the lighter hydrogen-rich fractions of the crude oil escaped from the rock. Simple treatment of the sands with hot water or steam will release the bitumen, to which hydrogen can be added to produce crude oil. The bitumen does contain 4 to 6 percent sulfur, however, which must be removed. The largest known reserve of this material is in the Athabasca tar sand in northern Alberta (Fig. 6-3). This rock layer covers an area of over 30,000 square miles, reaches thicknesses of over 200 feet, and is near enough to the surface in many places to be strip mined. In addition to the Athabasca tar sands, there are many smaller areas of tar sands scattered throughout western North America, as well as many deposits of "heavy oil," which, though still fluid, cannot be recovered readily by conventional pumping.

Oil shales, which are fine-grained sedimentary rocks rich in organic material, represent a larger technologic challenge. In contrast to the residual bitumen in tar sands, oil shales contain organic material that has yet to be converted to oil or gas. To do this requires temperatures of over 900°F and extensive processing. Most good oil shales will yield one-half to three-quarters of a barrel of oil per ton of shale. The largest, well-known reserves of oil shale in the world are in the Green River formation in the Western United States (Fig. 6-3) where layers considered minable range up to 100 feet in thickness. The Green River shales were deposited in a large, saline lake about 40 million years ago. Processing of oil shales requires about 100 gallons of water for each gallon of oil recovered, a large demand for the arid country in which our North American oil shales are found. Tremendous amounts of waste, with a volume about 15 percent greater than that of the original shale, will also plague attempts to mine the oil shales.

Future Supplies of the Fossil Fuels Estimates of the world's original fossil fuel reserves have been made by many investigators. According to one of the most widely accepted of these estimates (Table 6-3), we have consumed only 5 to 20 percent of the oil, natural gas, and coal, and essentially none of the oil shale and tar sand. Nevertheless, it is estimated that the rapidly increasing rate of oil and gas consumption will cause exhaustion of these supplies in less than a century. It seems clear, therefore, that oil and gas must be replaced by other energy sources in the very near future, particularly if we are to conserve them as hydrocarbon raw material for the chemical industry. Presently estimated tar sand and oil shale reserves are not large enough to change this estimate. Coal supplies, however, are thought to be adequate for several centuries, which suggests that it is this fossil fuel that will see us through the transition to large-scale use of other forms of energy such as nuclear energy.

NUCLEAR FUELS
Energy, partly in the form of heat, can be derived from fission and fusion. These are nuclear reactions, in that they involve the nucleus,* or core, of an

---

*Atoms consist of a central nucleus surrounded by a cloud of negatively charged electrons. The nucleus consists of positively charged protons and uncharged neutrons. Although the number of electrons and protons is fixed and characteristic for each element, the number of neutrons can vary in a small range to produce isotopes of each element. The total number of protons and neutrons in the nucleus of an isotope is indicated by a superscript such as uranium 235 (^{235}U) or uranium 238 (^{238}U). Since ^{238}U contains three more neutrons than ^{235}U, it is a heavy isotope of uranium.

Table 6-3 Approximate magnitude and energy contents (thermal joules $\times 10^{-21}$) of the world's original supply of fossil fuels recoverable under present conditions

FUEL	QUANTITY	ENERGY CONTENT	PERCENT
Coal and lignite	2.35×10^{12} metric tons	53.2	63.78
Petroleum liquids	$2,400 \times 10^9$ bbls	14.2	17.03
Natural gas	$12,000 \times 10^{12}$ ft^3	13.1	15.71
Tar sand oil	300×10^9 bbls	1.8	2.16
Shale oil	190×10^9 bbls	1.1	1.32

SOURCE: Hubbert (1973).

atom. In fission reactions, the nucleus of a single atom is split into several parts, whereas in fusion reactions, two nuclei combine to yield a new nucleus. Large elements such as uranium and thorium can provide fission energy whereas small elements like lithium and hydrogen can provide fusion energy.

Fission Energy Uranium, the principal source of fission energy, is found in three major types of deposits. In the United States, the most common type is the sandstone uranium deposit in which carnotite (Appendix I) fills the pore space in sandstones. These deposits are widespread in New Mexico, Colorado, Utah, Wyoming, and Texas. At Elliot Lake, Ontario, uraninite is found as small placer (Chap. 10) grains in extensive conglomerates that were deposited by streams more than 2 billion years ago. In Saskatchewan and in northern Australia, uranium minerals are found in veins in Precambrian rocks. In most of these deposits the present lower minable grade is 0.07 percent uranium, though material with lower grades is abundant. For instance, the extensive gold-bearing conglomerates of the Witwatersrand basin in South Africa (Chap. 9) contain about 0.03 to 0.06 percent uranium.

Including much of this lower-grade material, the U.S. Geological Survey recently estimated that worldwide reserves contain about 1.4 million tons of uranium. Several times that amount is thought to be present in lower-grade concentrations, especially in phosphate deposits (Chap. 5) and black, hydrocarbon-rich sedimentary rocks such as the Chattanooga shale, which underlies large areas of the southern Appalachians.

Thorium is found largely in heavy and resistant minerals such as monazite (Appendix I), which are concentrated along with other minerals in placer deposits and modern beach sands (Chap. 10). World monazite resources are large, though very few deposits can be mined for their thorium content alone.

Just how useful these fissionable resources are and how long they might last depends strongly on the state of nuclear reactor technology. Of all the natural isotopes, only ^{235}U will undergo fission spontaneously in the natural environment. However ^{235}U constitutes only about 0.7 percent of total uranium, the remainder consisting largely of ^{238}U, which must be bombarded by fission particles before it undergoes fission. The most abundant isotope of thorium, ^{232}Th, is similar to ^{238}U in this respect.

Thus, most present nuclear reactors use ^{235}U as their principal fuel and depend on products of the first fission reaction to cause new fission reactions which sustain the chain. Of course, the energy and number of such fission products must be controlled in order for the reactor to function as a power plant rather than a bomb and on this point technology can take several paths. In the United States, the uranium is processed to enrich it in ^{235}U. This fuel is then used in reactors in which the fission process is moderated by ordinary water. In Canada, unenriched uranium is used as reactor fuel because the process is moderated by special deuterium-rich, heavy water. In both of these systems, however, it is only ^{235}U that undergoes fission, and, therefore, less than 1 percent of the original uranium ore is actually used to generate power. Thus, these reactors are fittingly known as "burners."

In order to use uranium (and thorium) more efficiently, it is necessary to design reactors in which the energy of the fission products is high enough to convert significant quantities of ^{238}U and ^{232}Th to plutonium 239 (^{239}Pu) and ^{233}U, respectively, both of which are spontaneously fissionable just like ^{235}U. If such a reactor produces more fissionable material than it consumes, it is a "breeder" reactor. At present, however, breeder-reactor technology is not as highly developed as is burner-reactor technology, and burner reactors continue to be installed. Unless these burner reactors are replaced by breeder reactors in the near future, it is generally accepted that we will deplete our fissionable resources in just a few decades. With breeder reactors in use, however, we could obtain much more energy from uranium and, therefore, we could mine lower-grade ores. This would, in turn, increase the potential of fission reactions to supply our energy needs for several centuries.

One of the major problems besetting the large-scale use of fission power is the disposal of nuclear wastes. The unfortunate fact is that when uranium and thorium undergo fission, they produce products that are unstable themselves and will undergo further nuclear decay reactions. Many of these products, such as strontium 90 (^{90}Sr), are especially dangerous because they will concentrate in parts of the body and yield dangerously high levels of radiation. To allow these radioactive wastes to decay to safely low levels, we must isolate them from the natural environment for at least 250,000 years. In the United States alone, about 100 million gallons of such waste is in storage awaiting a decision on where it can be stored safely for such a long time.

Among the possibilities are: (1) storage at the base of the Antarctic ice sheet, (2) storage in rock chambers at great depth, (3) injection of the wastes into deep wells, and (4) rocket transport of the wastes to the sun or space. The most reasonable of these options seems to be storage in artificial caverns in layers of evaporite salt (Chap. 5). Such caverns, being made of relatively plastic salt, do not have many fractures that might allow water to enter or radioactive wastes to escape. Salt domes are a second choice because their upward movement over long time periods may make them less stable.

Fusion Energy Fusion reactors depend on more advanced (and still unperfected) technology than do fission reactors. The main problem is in the design of a container in which the exceedingly high-temperature fusion reaction can be sustained. At present, the deuterium-tritium* reaction is considered most accessible from a technological standpoint. However, tritium for this reaction must be derived from lithium ores (Chap. 4). M. K. Hubbert of the U.S. Geological Survey estimated that natural lithium ore abundance will limit the energy contribution of this reaction, if it is ever feasible, to an amount about equal to that of all the fossil fuels. Other fusion reactions, especially the deuterium-deuterium reaction, which would use ocean water as a deuterium source, could yield almost infinite amounts of

---

*Deuterium and tritium are isotopes of hydrogen containing zero and one neutrons in their respective nuclei.

64 OUR FINITE MINERAL RESOURCES

energy. For instance, Hubbert has calculated that the deuterium in 1 cubic kilometer of sea water could provide almost as much energy as all the oil in the world. All fusion reactors would share the advantage (over fission reactors) of producing very little radioactive waste.

GEOTHERMAL POWER

As you know, temperature increases downward in the earth. In the core of the earth, in fact, temperatures are thought to exceed 7000°F. Although all this heat is a tremendous potential reservoir of energy, we are limited at present to the heat in the outer 10,000 feet or so of the earth's crust. In this outer shell, temperatures are relatively low, and circulating water usually reaches the necessary minimum (to produce sufficient steam in the generating system) temperature of about 360°F in small areas of recent volcanic activity such as New Zealand, Japan, Italy, Iceland, Mexico, and California (Fig. 6-5). Here, wells tap steam to drive electric generators. Such natural geothermal systems are much less efficient than other methods of power generation. Most are gradually depleted of water and heat and have a useful lifetime of about 25 to 50 years.

Because of the limited global distribution of high-temperature natural geothermal systems, other ways have been proposed for using the more widespread, lower-temperature geothermal energy in the upper part of the earth's crust. Possibilities involve the use of heat-exchanger fluids that boil at lower temperatures than does water, and even explosion of a nuclear device

Figure 6-5 Power Plant Number 3 in The Geysers geothermal field, Sonoma County, California. This plant, which was activated in 1971, is fed by the steam wells in the foreground and has a capacity of 110,000 kilowatts. *(From California Division of Mines and Geology.)*

at depth and subsequent use of the artificially heated rock and water. The potential for geothermal energy to make a truly significant contribution to our energy resources on a worldwide scale depends upon technological developments such as these.

CONCLUSIONS

During the last few centuries, the major source of energy for the world has evolved from wood to coal and finally to oil and natural gas. The extent to which our industrial complex has become dependent on oil and gas has given these commodities an air of permanence that their reserves do not permit. We have forgotten that the evolution must continue. On the basis of the abundances of the mineral energy resources, it seems clear that larger fractions of our future energy requirements will be filled by nuclear energy, first as fission energy and then, perhaps, as fusion energy. With the exception of deuterium-deuterium fusion reactions, it seems likely that we are only several hundred years away from the exhaustion of our mineral energy resources. This impending shortage of energy is all the more serious because it will coincide with the depletion of most of our higher-grade metal resources. Thus, just at the time when we most need extra energy to process low-grade ore, we may not have it.

To fill this nearing gap, we must develop methods of using solar energy on a large scale. We must hope, also, as our technology evolves toward the economic use of solar energy, that such use does not depend upon metals or compounds already approaching exhaustion.

SEVEN

THE FRAMEWORK OF OUR ECONOMY—STRUCTURAL METALS

INTRODUCTION

Iron and aluminum are the basic materials for construction of everything from ships to spoons. Whereas iron has been used by man for over 3,000 years, however, aluminum attained real economic importance only during the twentieth century. Annual world steel production is even now more than 50 times greater than aluminum production, which clearly reflects the continued dominance of iron. This dominance centers on two facts. First, iron can be obtained simply by heating iron ores, whereas aluminum production requires large amounts of electrical energy. Second, because many iron ores are higher grade than aluminum ores, the yield of metal per ton of ore is greater. Thus, aluminum costs about 30 cents a pound and steel costs 10 cents a pound. Even at the same cost, steel would be preferred for many uses because it is stronger, though heavier, and because it can be combined (alloyed) with many other metals to produce steels with a wide variety of properties (Table 7-1).

Steelmaking, and to a lesser extent aluminum making, are such large-scale industries that their needs for other raw materials are tremendous. Coke and limestone are major necessities for steelmaking, as you probably learned in your first geography lesson. Coke, which is made by heating coal

Table 7-1 Elements commonly combined with iron to form steels and alloys with special properties

METAL	PROPERTIES CONFERRED TO STEEL	APPROXIMATE PERCENT OF WORLD CONSUMPTION USED IN STEEL INDUSTRY
Chromium	Corrosion resistance, high-temperature strength	55
Cobalt	High-temperature strength, permanent-magnet steels	50
Copper	Machinability, corrosion resistance	minor
Lead	Machinability	minor
Magnesium	Strength, ductility	2
Manganese	Strength, toughness, abrasion resistance	90
Molybdenum	Strength, ductility, shock resistance	90
Nickel	Toughness, corrosion resistance	60
Niobium	Heat resistance, strength	70
Silicon	Corrosion, heat- and wear-resistance	80
Tungsten	High-temperature strength, hardness	25
Vanadium	Strength, ductility, resiliency	90

SOURCE: Estimated from information in U.S. Bureau of Mines publications.

to drive off hydrogen, represents almost one-quarter of U.S. and Canadian coal consumption. Some elements such as manganese, chromium, and fluorine are so essential to steel and/or aluminum making that their availability could control the continued health of these industries. Likewise, since the steel industry is the major user of many of the elements listed in Table 7-1, the health of the steel industry is reflected by much of the mining industry.

IRON AND STEEL

Steelmaking Present-day steelmaking is a two-stage process. First, the iron ore is smelted with limestone and coke in a blast furnace where much of the impurities combine with lime to form slag. During the smelting operation, coke reacts with iron ore (which is iron oxide) to produce carbon dioxide and pig iron containing 3 to 4 percent carbon. The pig iron is then transferred to another furnace in which it is converted to steel by removing most of the carbon and as much as possible of other impurities. At this stage, the iron can be mixed with elements such as manganese or chromium (Table 7-1) to provide special steels with the capacity to withstand abrasion, heat, corrosion, shock, or strain. In modern steel plants, all parts of the process are closely connected to minimize fuel requirements. Nevertheless, a single-stage steelmaking process, which would be cheaper, remains a major goal for research.

Prior to about 1940, it was felt that steel could be produced economically only by countries in which there was a favorably located combination of iron ore, limestone, and coal. The development of larger and more efficient ocean transportation systems, however, has rendered this concept invalid. The urge to establish a national steel industry is always strongest where iron ore is available but coal is scarce and, as shown by both Mexico and Sweden, this can be done successfully. Many people in the steel business feel that the Middle Eastern countries, especially Saudi Arabia, may soon use their excess natural gas to establish highly competitive steel mills.

The undisputed champion, however, is Japan, which has become the world's third largest producer of steel and a major steel exporter although it imports about 95 percent of the iron ore, 80 percent of the coking coal, and 30 percent of the scrap iron used to make this steel. In 1970 alone, Japan imported over 100 million tons of iron ore, over 50 million tons of coking coal, and over 6 million tons of iron scrap. The cost of these materials, as well as the cost of imported fuel for the steel mills, amounted to almost $3 billion. All these expenses were met by the export of less than 20 million tons of steel products, leaving a sizable surplus of steel for domestic use. The advent of more expensive energy may limit Japan's continued success as a steel manufacturer, however.

Iron Deposits How firm is the base on which the steel industry rests? Is there enough iron ore to sustain our present annual world steel consumption of almost 800 million tons? The answer is a definite yes (Table 7-2). We are endowed with at least enough iron ore to last a century and probably much longer, largely because of one type of deposit, the sedimentary iron formations. These consist of layered sedimentary (or metamorphosed sedimentary) rock containing at least 15 percent iron. The iron can be in magnetite, hematite, limonite, pyrite, siderite, or other minerals (Appendix I), which are usually concentrated in thin iron-rich layers that alternate with iron-poor layers of valueless chert, a water-rich form of silica (SiO_2). Most sediments of this type are ten to hundreds of feet thick and are found in the partly eroded remains of old sedimentary basins that were originally many hundreds of square miles in extent (Fig. 7-1). The sedimentary iron formations in both the Lake Superior (U.S.) and Hammersley Range (Australia) areas (Fig. 7-2) were deposited in basins over 300 by 100 miles in extent. The Labrador trough iron formations extend for over 500 miles through eastern Canada. Many of the sediments in these old basins have been strongly folded and in many places they are deeply buried. Only where they are near the surface can they be mined under present conditions.

The basins in which these iron-rich sediments were deposited seem to have been present only in Precambrian rocks with an age of 1.7 billion years or more. The exact reason for this is not well known but is thought to relate to

Table 7-2 World iron ore resources (millions of metric tons). As discussed in the text, iron ore can range from about 20 to 65 percent iron. Thus, the content of iron metal in these tonnages of iron ore varies from place to place.

	PRESENTLY ECONOMIC KNOWN RESERVES	LOWER-GRADE ORES
United States	9,000	92,000
Canada and Mexico	36,000	90,000
South America	34,000	60,000
Europe	21,000	13,000
Africa	7,000	24,000
Australia-New Zealand	17,000	large
Asia	17,000	55,000
USSR	111,000	193,000
TOTAL	252,000	527,000+

SOURCE: United Nations, 1970.

the possibility that the composition of the atmosphere at that time was low enough in oxygen to permit iron to be transported to sedimentary basins more easily.

Although the enormous volumes of sedimentary iron formation constitute a very large iron resource, only parts of the iron formations are minable at present. For one thing, the steel industry is geared to use iron oxide minerals because they have a relatively high iron content (Appendix I). Thus, iron carbonate or iron sulfide ores, which contain less iron, must be converted to oxide form before they can be treated in most conventional steel mills. This, of course, inserts an extra step in the process, making it more expensive. In fact, only one North American mining area, Ducktown, Tennessee, produces significant amounts of iron from nonoxide ores, and this is a by-product in the roasting of sulfide ore. Only those parts of iron formation containing iron as iron oxides, therefore, are of widespread economic interest.

A further complication results from the fact that iron formation in its original, unweathered condition is relatively difficult to mine and process because over half of the rock consists of very hard chert and other impurities. Where rain water has percolated downward through this hard formation, known as taconite, the chert has been gradually leached out, leaving large tonnages of relatively soft, enriched ores containing 50 to 70 percent iron as iron oxide. In the Lake Superior region, which has supplied over 60 percent of domestic U.S. iron ore, these high-grade surface ores were the mainstay of mining for over 50 years until they were largely depleted in the 1950s. At this time, new methods had to be developed to exploit the hard iron oxide-bearing taconite ores. It was necessary to crush the taconite, concentrate the iron oxide minerals, and press the concentrate into pellets (small, half-inch balls of concentrate and clay averaging about 60 to 68 percent iron). Because of their even size, these pellets (Fig. 7-1) are so much easier to transport and cheaper to melt in blast furnaces that they have gained a competitive edge

Figure 7-1 (a) Iron formation at the Soudan Mine in Minnesota consists of tightly folded alternating layers of iron oxide minerals (black) and silica (white). (b) Pellets, seen here on a traveling grate at the Empire mine in Michigan, are the concentrate formed by separating silica from the iron formation. *(From* Mining Engineering.*)*

over the soft, higher-grade ores. In fact, pellets already account for almost one-fifth of world iron ore production, and it is expected that their importance will continue to grow rapidly.

The development of taconite mining and pelletizing greatly enlarged the presently economic iron ore reserves available in countries such as Canada and the United States. Other countries, such as Australia and Brazil (Fig. 7-2) are not yet at this stage in their resource consumption, however, and still have large reserves of high-grade weathered ores to be mined. Interestingly, Australia banned export of iron in 1938 when it appeared that domestic reserves were inadequate for its short-term needs. This effectively stopped exploration for iron ore in Australia until 1960 when the embargo was lifted. Within 2 years, large discoveries were made in Western Australia and within 12 years Australia had become the world's leading exporter of iron ore exploiting high-grade, weathered iron formation ores with 60 to 65 percent iron. Well-established reserves of Australian iron ores containing over 60 percent iron totalled more than 5 billion tons in 1970, with enormous tonnages of lower-grade ore remaining to be assessed.

There are other types of iron ore deposits, some of which are very large. Many of these, such as the Kiruna ores of Sweden and the now-exhausted Cornwall, Pennsylvania deposit (Fig. 7-2), were apparently formed by igneous processes. Others, such as the Red Mountain iron ores, which originally supplied the Birmingham, Alabama steel mills, are sedimentary ores formed under conditions more like those of the present. Including all these ores, the U.S. Geological Survey estimates that presently minable deposits contain at least 80 billion tons of iron metal and that lower-grade deposits throughout the world contain an equal amount. By comparison, we are mining (worldwide) almost 900 million metric tons of iron ore each year (containing more than 500 million tons of iron). Thus, there is ample time to contemplate what should be done when these ores are exhausted.

Environmental Problems with Iron and Steel Production The only really significant difficulty to arise from the mining and beneficiation of iron ore is the long standing controversy over the tailings (Chap. 2) that are being dumped into Lake Superior by a company that pioneered taconite mining in Minnesota. Usually, tailings are dumped on land in large ponds that can be difficult to maintain and expensive to revegetate. In an effort to do away with its tailings, the company discharged them into Lake Superior. Subsequent controversy has developed over whether these tailings, which consist of chemically inert rock fragments, are in fact sinking harmlessly to the bottom of the lake or whether they are affecting water supplies and aquatic life.

The real environmental ogre in the steel industry, however, has been the blast furnace. Because large amounts of oxygen are needed to maintain high temperatures in the furnace, air is forced through the ore-coke-limestone mixture and the resulting gases that are vented into the atmosphere carry large volumes of fine particles which cause air pollution. The advantage of the cohesive pellets in minimizing the emission of particles from blast furnaces is obvious. In addition, major efforts, estimated by American steel

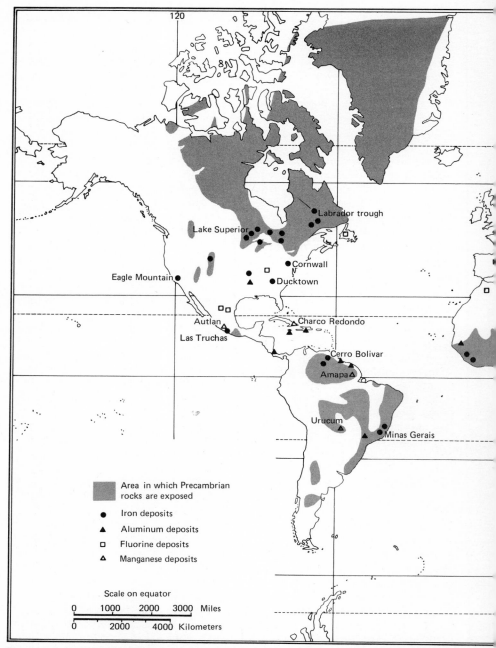

Figure 7-2 Distribution of major iron, aluminum, fluorine, and manganese deposits in the world. Note that almost all the iron deposits are in Precambrian rock (shaded areas).

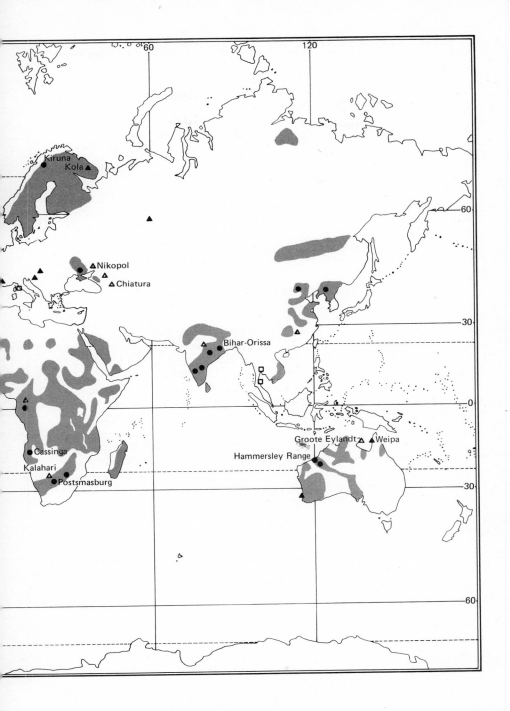

companies to cost about $300 million in 1972, are being made to curtail emission of gases and other particles.

ALUMINUM

Aluminum metal is often referred to as "packaged power" because of the tremendous amounts of energy necessary to convert the ore to the metal. Whereas many copper ores contain about $5 to $10 worth of copper in each ton, the most common ore of aluminum (bauxite) contains 20 percent or more aluminum with a value of about $120 on today's markets. The value difference results from the cost of processing. Aluminum production is a two-stage process. First, bauxite is leached with hot caustic soda to remove impurities such as silicon, titanium, iron, and water, leaving pure aluminum oxide (alumina). The alumina is then mixed with molten sodium aluminum fluoride (cryolite) at a temperature of about 1800°F, an electric current is passed through the mixture, and molten aluminum separates from the alumina. About 5 to 7 tons of bauxite, 1,000 pounds of carbon (largely from petroleum), over 50 pounds of fluorine, and 7 to 8 kilowatt-hours of electric energy are required to produce 1 ton of aluminum. The necessity for such large amounts of electricity has made it more practical to take the bauxite to sources of cheap power than to generate power at the mines. Thus, several countries rich in hydroelectric power, notably Canada and Norway, have become major producers of aluminum metal although they have no presently minable bauxite. Here again is an opportunity for the oil-rich Middle East to use its excess power.

Bauxite is formed by intense weathering of aluminum-rich rocks and minerals exposed at the earth's surface. Under suitable conditions including abundant rainfall, good drainage, heavy vegetation, and high temperatures, most of the elements present in these rocks and minerals will be removed in solution, leaving aluminum, oxygen, and water combined as hydrous aluminum oxides. Obviously, any natural aluminum concentrating process will work best on a rock with a high original aluminum content such as feldspar-rich igneous rocks or clays. In some areas, such as Jamaica and southern Hispañiola, bauxite is found on limestone bedrock where clays that collected in depressions and sinkholes on the limestone surface have been intensely leached. Although most bauxite deposits are found in today's tropics (Fig. 7-2), a few, such as those in France and Arkansas, are preserved from earlier times when climates in those areas were warmer. Because bauxite deposits are formed by surface processes, they are easily removed by erosion unless buried under a protective cover of younger sediments as in northern South America and Arkansas (Fig. 7-3).

The aluminum resource picture is a relatively cheerful one. Known deposits in Jamaica, Surinam, Guyana, Guinea, Ghana, Australia and other countries are estimated to contain 2 to 5 billion tons of metal in comparison to present annual world production of 10 to 20 million tons. In addition, there are huge tonnages of rocks and minerals containing somewhat less aluminum than bauxite. One of the few areas where such rock is mined today is in

Figure 7-3 Strip mining of bauxite at MacKenzie, Guyana. Surficial Pleistocene sand is removed by bucket-wheel excavator (upper left). The tougher, underlying sandy clays and bauxite (bottom layer) are mined by draglines (right and lower left). Mining proceeds from right to left, and waste from new stripping is piled in the mined-out areas on the right. *(Courtesy of Thomas L. Kesler.)*

the USSR, where high-grade bauxite is scarce. Here, the feldspar-rich igneous rock, nepheline syenite, is mined in the Kola peninsula and the Chulim river areas, and alunite, an aluminum-bearing sulfate, is mined at Kirovobad. Several countries have successfully recovered aluminum from shale, clay, and other nonbauxite sources during national supply emergencies, however, and it seems likely that technological developments will make these resources more widely usable in the future.

MANGANESE

Manganese is as essential as steel, no more, no less. Between 13 and 34 pounds of manganese are needed to produce one ton of steel, and over 90 percent of world manganese production is used in the steel industry. Manganese is essential because it is used to remove sulfur and oxygen from pig iron and as an additive to increase the strength of steel. Presently minable world manganese reserves are very large, in the range of 2 billion tons of manganese metal. This is enough to supply present annual world manganese consumption of about 10 million tons for many years to come. In addition, we can fall back later on abundant low-grade manganese resources on land, and unquantified, but large, resources of manganese-rich concretions that litter much of the sea floor (Chap. 10).

Unfortunately for many major steel producing nations, the distribution of manganese does not conform to that of iron ore. The United States, Canada, and Sweden, for instance, have essentially no economic manganese resources. Australia, India, and the USSR are relatively well supplied with both commodities, but only the USSR has the necessary coal supplies to capitalize on this coincidence and form a strong domestic steel industry.

Most minable manganese deposits consist of complex mixtures of manganese oxides (Appendix I) and have an overall content of 25 to 40 percent manganese. Many large manganese deposits, such as the enormous Chiatura and Nikopol deposits in the USSR and the Kalahari deposits in South Africa (Fig. 7-2), seem to have formed as sedimentary accumulations in shales and limestones. Smaller deposits, such as Autlan in Mexico and Charco Redondo in Cuba, may have formed in lakes and shallow seas fed by volcanic hot springs. As with iron, there are many zones of manganese-rich rock, such as the huge Moanda deposits in Gabon, that are not originally ore grade but which yield rich manganese oxide ores at and near the earth's surface when impurities are leached out of the rock during weathering.

CHROMIUM

Although most mineral resources are scattered erratically around the world, chromium makes most others seem evenly distributed. Over 95 percent of the identified chromite (Appendix I) resources in the world are in Rhodesia and the Union of South Africa. To make matters more extreme, Rhodesia controls about 85 percent of the world's high-chromium (31 percent or more chromium) chromite. The high-chromium chromite is used largely to produce stainless steel, which accounts for over half of world chromium consumption. Low-chromium chromite, on which South Africa has the monopoly, is used most widely to make refractory materials, such as linings for metallurgical furnaces, where chemically inert compounds with high melting temperatures are needed.

The large chromite reserves of Rhodesia and South Africa are found in enormous bodies of magnesium-iron-rich igneous rock that is layered very much like a sediment. This layering apparently formed as the magma crystallized and the early-formed crystals settled to the bottom of the magma enclosure. Chromite was among these minerals and somehow it became concentrated in certain layers, most of which range from an inch to a few feet in thickness. The Bushveld complex (Fig. 9-1), as the chromite-bearing rock unit in South Africa is called, is a saucer-shaped body that covers an area of almost 300 by 150 miles and is about 5 miles thick. (Remember the Bushveld complex, for you will hear more about it in Chap. 9.) The Great Dike, which is one of the ore hosts in Rhodesia, is over 300 miles long and 3 to 6 miles wide. The Selukwe complex, also in Rhodesia, is a smaller but similar rock unit that has been strongly folded. Presently minable near-surface parts of the chromite layers in the Bushveld, Great Dike, and Selukwe are estimated to contain over 500 million tons of chromium in comparison to present annual world chromium consumption of about 2 million tons. Clearly, it will be problems of distribution (Chap. 1) rather than geological scarcity that will cause any shortages of chromite in world markets.

FLUORINE

Fluorine is a unique mineral commodity in that it is so closely tied to both the steel and aluminum industries. Of the nearly 200,000 tons of fluorine consumed each year in the United States, about one-third is used as a flux to lower the melting temperature of furnace charges during steelmaking and about one-fifth goes into the molten bath in which alumina is converted to aluminum.

Fluorine is found most commonly in nature as the mineral fluorite, a calcium fluoride (Appendix I), which is known as fluorspar in industry. Most of the world's important fluorine deposits are veins, masses, and layers of nearly pure fluorspar in limestone. Most deposits have grades of about 20 to 40 percent fluorine with a value of about $10 to $40 per ton of ore. Individual deposits are not particularly large, rarely amounting to a few million tons of ore. In some deposits, such as those in the Parral district of Mexico, fluorite is a byproduct of lead and zinc mining. The fluorspar in all these deposits seems to have been deposited by hydrothermal solutions (Chap. 8) circulating through fractures and holes in the host rock. In at least some of the deposits, such as those in northern-most Coahuila, Mexico, it is clear that the water came from nearby intrusive igneous rocks.

Present world fluorine production of about 2.5 million tons comes largely from Mexico, the USSR, and Europe (Fig. 7-2). Total identified world reserves were recently estimated by the U.S. Geological Survey to be 30 million tons, of which about one-third is in North America. About twice that amount is thought to await discovery and higher prices. Thus, unless major new fluorine sources are discovered or new methods of steel and aluminum production are developed, fluorine may become the weak link in the framework metal production chain.

EIGHT

EXTREME NATURAL CONCENTRATIONS—THE SCARCE METALS

INTRODUCTION
In addition to the structural elements discussed in Chapter 7 there are many metals, ranging from antimony to zinc, which have an essential niche in our industrialized society. All these metals are relatively scarce in the earth's crust and require natural concentrations of 100 to 1,000 times to qualify as ore deposits (Chap. 1). Some, such as copper, molybdenum, nickel, lead, zinc, mercury, tin, and tungsten, tend to form deposits in which one of them is the principal element of value. Many others, such as bismuth, cadmium, cobalt, indium, rhenium, and thallium, are usually found in small but recoverable amounts in deposits of more abundant metals. The distinction between the two groups is not absolute. Antimony and arsenic, for instance, can be recovered from copper, mercury, lead, and zinc ores but are also found alone in important deposits of their own. In the following sections, we will confine our attention to the scarce metals that are found largely in deposits of their own.

COPPER
Copper is valued for its combination of high thermal and electrical conductivity, the ease with which it can be shaped, and its unusual red color. Approximately 7.5 million tons of newly mined copper were used in 1973 in

wires, pipes, alloys (brass, bronze), and other products. Aluminum can substitute for copper in many uses and competition between the two metals is strong. Copper prices have more than tripled during the last few years to over $1 a pound (in 1974). This rise has been stimulated by copper shortages arising from expropriations (Chap. 1) as well as the inability of new production to meet growing demand. Although copper is recovered from a wide variety of deposits, two types, the porphyry copper deposits and the sedimentary copper deposits, account for over two-thirds of world production (Table 8-1). Massive sulfide deposits and nickel-copper deposits account for most of the remainder.

Porphyry Copper Deposits Porphyry copper deposits are the "discount stores" of the mining business. Typical porphyry copper ore grades about 0.4 to 1.0 percent copper and contains only 1 to 4 percent of the copper minerals chalcopyrite and chalcocite (Appendix I) disseminated through an otherwise worthless rock. Such a low metal content requires that the ore be mined in large volumes in order to maintain high rates of copper production. This, of course, necessitates a very large investment in mining and processing equipment and correspondingly tight cost control for profitable operation. One of the champions along these lines is the Sierrita mine in Arizona. It required an initial investment of $165 million and handles, each day, about 140,000 tons of waste and 80,000 tons of ore (averaging 0.29 percent copper and 0.028 percent molybdenum). In the largest deposits, such as the Bingham mine in Utah and the Panguna mine in the Solomon Islands, almost half a million tons of ore and waste are moved daily.

Porphyry copper deposits derive their name from their close association with porphyritic, intrusive, igneous rocks, which consist of large mineral grains in a matrix of smaller ones (Fig. 8-1). The deposits are thought to have formed where a feldspar-rich magma, injected upward to within a mile or so of the surface, crystallized fairly quickly expelling the water it contained to form a hot, water-rich fluid known as a hydrothermal solution. (The release of a hydrothermal solution is roughly analogous to the expulsion of dissolved carbon dioxide when a carbonated drink is frozen.) Hydrothermal solutions

Table 8-1 Approximate contributions of the four major types of copper deposits to total world production (exclusive of Communist countries for which production statistics are scarce)

DEPOSIT TYPE	LOCATIONS OF MAJOR DEPOSITS	COPPER METAL (tons)
Porphyry copper	US, Canada, Chile, Peru, South Pacific	3,000,000
Sedimentary copper	Zambia, Zaire	1,300,000
Massive sulfide	Canada, US, Japan	400,000
Copper-nickel	Canada	220,000

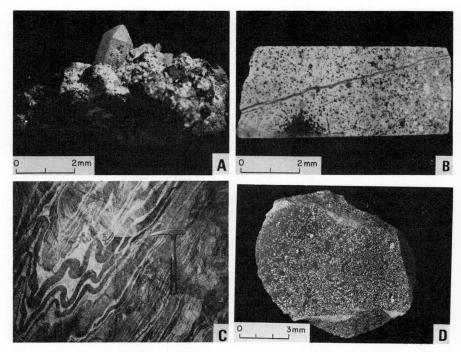

Figure 8-1 Deposits of the scarce metals were formed in many ways, some of which can be determined from examining the texture of the ore. The minerals of sample A, from the Mochito silver-lead-zinc mine in Honduras, grew into an open space (note the well-formed quartz crystal). Sample B, from the Cerro Colorado porphyry copper deposit in Panama, is an igneous rock with quartz (gray) and plagioclase (white) phenocrysts and disseminated chalcopyrite (black) cut by a small vein of anhydrite. Sample C, from the Sullivan lead-zinc mine in British Columbia, shows alternating layers of galena (white) and sphalerite (dark) that were presumably deposited as sediments and later folded. Sample D, from the Alexo nickel deposit in Ontario, consists of small blebs of the nickel-bearing sulfide minerals, pyrrhotite and pentlandite filling the spaces between larger grains of silicate minerals. These sulfide minerals presumably accumulated as molten sulfide material as the magma crystallized and do indeed increase in abundance downward (to the right in this photo).

are believed to be relatively acidic when they are expelled by the magma and to contain dissolved metals, such as copper, which did not enter into the minerals of the magma as it crystallized. Thus, the hydrothermal solution could leach and alter the surrounding rock and, in some cases, deposit copper minerals.

Most of the economically minable porphyry copper deposits are found in modern island arc volcanic zones such as New Guinea and the Philippines and in the geologically young mountain chains such as the Andes and the Rockies (Fig. 8-2). A few older, roughly similar deposits have been found in the Appalachians and the Precambrian rocks of Canada.

Sedimentary Copper Deposits Most sedimentary copper deposits consist of disseminations and layers of copper sulfide minerals (Appendix I) that form extensive tabular bodies more or less parallel to the layering of the enclosing sediments. Unlike porphyry copper deposits, most sedimentary deposits have higher grades, in the range of 1 to 8 percent copper. Because sedimentary copper deposits were formed by relatively widespread sedimentary processes, copper-bearing layers can be continuous for tens or even hundreds of miles, although only parts of such layers are usually of sufficient thickness and grade to permit mining. Almost everyone agrees that the copper in these deposits was precipitated from water. However, the copper content of most present-day surface waters seems too low to do the job (Table 3-2). Possibly, extra copper was added to the water locally from hot springs or from weathering of nearby copper-rich rocks or even preexisting copper deposits. Perhaps, too, the available copper was efficiently concentrated by living and dead organic matter. Copper-rich groundwater could also have been involved in areas such as White Pine, Michigan, where the copper sulfide, chalcocite (Appendix I), is disseminated through a widespread Precambrian shale.

The largest of these sedimentary deposits, such as the enormous "Copper Belt" of Zambia and Zäire in Africa (Fig. 8-2), were formed about 700 million years ago. The extensive copper-rich Kupferschiefer beds were deposited in the 260-million-year-old Zechstein evaporite (Chap. 5) basin in northern Europe. Similar, though so far less productive, copper-rich deposits were formed throughout much of Oklahoma and adjacent states at about the same time.

Massive Sulfide Deposits Massive sulfide deposits, though of less importance to the present copper supply picture (Table 8-1), merit mention because of their contribution to the mining industry of eastern Canada. The deposits are roughly lens-shaped bodies of sulfide minerals containing iron, copper, zinc, and, in places, lead, that are found interlayered with volcanic rocks deposited in sea water. They are thought to represent peculiar accumulations of metal-rich mud formed at or near the mouth of submarine hot springs during lulls in volcanic activity (Chap. 10). Perhaps because volcanic rocks were deposited underwater in relative abundance in the early part of the earth's history, massive sulfide deposits are unusually abundant in volcanic rocks more than about 2.5 billion years old and are widespread in eastern Canada. Similar deposits, however, can be found in rocks of almost all ages. The important Kuroko deposits of Japan, for instance, are less than 60 million years old.

Most massive sulfide deposits have maximum dimensions of only a hundred meters or so, are less than 50 meters thick, and contain only 2 to 4 million tons of ore. A few, however, such as the Horne mine in Quebec and the Kidd Creek mine in Ontario (Fig. 2-1) reach sizes of 80 to over 150 million tons. Almost all massive sulfide deposits have copper grades of 1 to 5 percent and contain relatively large amounts of other metals, especially zinc.

Massive sulfide deposits are among the most elusive of exploration

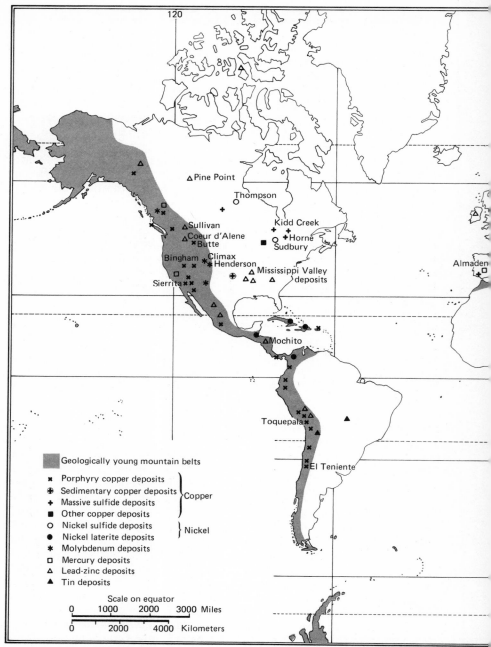

Figure 8-2 Distribution of major deposits of the scarce metals. Note the close association of porphyry copper and mercury deposits with the geologically young mountain belts (shaded). These mountain belts were formed during the last 200 million years or so by several tectonic events, not all of which were worldwide.

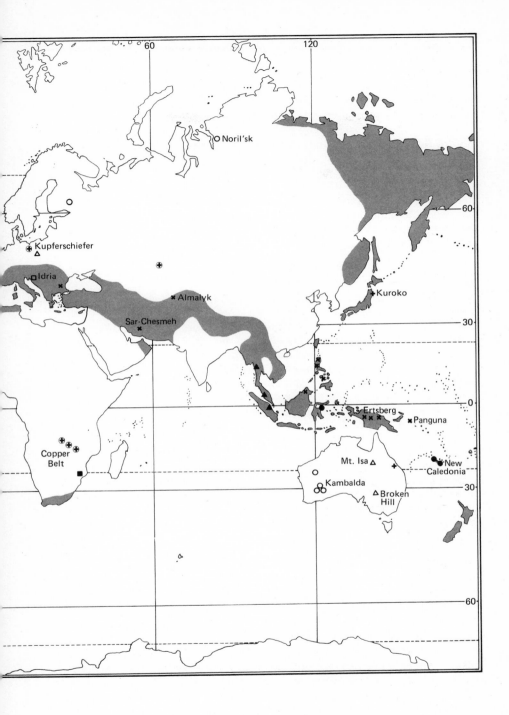

targets because they are so small. Even the small ones can be financial plums, however, because their high grade permits low-volume mining that requires comparatively small investments in equipment.

Future Supplies The outlook for future copper supplies is fairly optimistic, at least for our generation. Recent estimates by the U.S. Geological Survey suggest that about 350 million tons of copper are present in known deposits, most of which are minable at present. At least another 1,000 million tons or so are thought to be present in lower-grade conventional deposits, and in deposits that remain to be discovered. Such an amount could sustain us well into the twenty-first century, if copper prices remain high enough to stimulate exploration.

MOLYBDENUM

About half of world molybdenum production is a by-product from porphyry copper deposits; the other half comes from copper-free porphyry molybdenum deposits. In both types of deposits, the ore mineral is molybdenite (Appendix I). Molybdenum deposits are larger and more abundant on continents and are poorly developed in most island arc areas (Fig. 8-2). The really large porphyry molybdenum deposits, such as Climax and Henderson in Colorado, are found on the continent side of the porphyry copper deposits (Fig. 8-2). In relation to annual world molybdenum consumption of about 100,000 tons (most of which is used in steel), there is estimated to be over 30 million tons of molybdenum in known deposits that are presently economic or that will become economic in the next few decades as demand increases.

In addition to this rosy supply picture, it is comforting to know that molybdenum mining is being carried out in scenic mountain areas of Colorado with such care that it has brought awards from concerned environmentalists. The $200 million Henderson project, for example, includes a 10-mile tunnel below the continental divide to take ore from the underground mine to an area of lesser scenic appeal where it can be processed.

NICKEL

Nickel, like molybdenum, finds its greatest single use in the steel industry. Over 60 percent of world nickel consumption is used to provide steel with resistance to corrosion and heat. Nickel tends to concentrate in metal sulfide minerals (Appendix I) if enough sulfur is available. Where these nickel-bearing sulfide minerals have been concentrated sufficiently to provide nickel grades of about 1 percent or more, minable nickel deposits are the result. Where sulfur is deficient, however, nickel tends to concentrate (at levels of about 0.4 percent) in magnesium silicate minerals such as olivine. Olivine melts at such high temperatures that it is not economically feasible to remove this nickel, however. Fortunately, intense tropical weathering breaks down olivine and other magnesium silicate minerals to form nickel-bearing oxides and silicates in which the nickel is contained in more easily processed forms. These deposits, known as nickel laterite deposits, can have grades of

1 to 10 percent nickel, with the lower-grade ores being much more common. As might be expected, nickel laterites develop best on rock that is originally rich in nickel (0.1 to 0.3 percent). Such rocks, which are found in the earth's mantle, have been shoved up to the surface locally along deep fractures. One spot where this has happened, New Caledonia in the southwest Pacific, produces about one-fifth of the world's nickel (Fig. 8-2).

Nickel sulfide deposits, which are of greater present economic importance than the nickel laterite ores, appear to have formed when droplets of iron sulfides rich in nickel, and sometimes copper, separated from magnesium-rich magma and sank before much of the magma could crystallize, collecting at the bottom of the magma enclosure (Fig. 8-1). This process took place almost exclusively in rocks older than about 1.5 billion years, which limits most of the world's nickel sulfide deposits to older Precambrian rocks exposed in Canada, Africa, and Australia (Fig. 7-2). Although most nickel sulfide deposits such as those in the Eastern Goldfields area of Western Australia around Kambalda, contain only 1 to 35 million tons of ore, those found in the Thompson and Sudbury areas of Canada are so large that they provide about 40 percent of world production. The Thompson deposits, though strongly folded now, could be very large analogues of the Kambalda deposits.

The Sudbury deposits, however, are unique. For one thing, they are found at or near the base of a funnel-shaped intrusive rock that is much larger and poorer in magnesium than the rocks at Thompson and Kambalda. This huge intrusion at Sudbury is thought to have formed when a meteorite struck the area causing molten rock and nickel sulfides to rise up from the mantle. As far as we know, this combination of meteorite impact followed by magmatic intrusion is unique in the history of the earth, and, therefore, Canada will probably dominate the nickel industry for many years to come.

Reliable estimates of future nickel supplies are somewhat scarce, partly because of lack of information on reserves at Sudbury and partly because new discoveries are being made fairly rapidly at present. Between 50 and 100 million tons of nickel are thought to be in known sulfide and laterite deposits that are minable at or near present prices ($1 to $2 a pound). Enormous amounts of lower-grade material await the technological breakthroughs necessary to make them minable. At annual consumption rates of 400,000 to 500,000 tons of nickel, these reserves are comforting.

LEAD AND ZINC

Although lead and zinc find widely different uses in industry, they occur together so commonly in mineral deposits that they can be considered here in a single section. Both elements are found in a wide variety of deposits including the massive sulfide deposits mentioned previously. All these deposits contain veins and masses of galena and sphalerite (Appendix I) that show evidence of having been deposited from hydrothermal solutions. Some deposits, such as those surrounding the porphyry copper deposit at Bingham, Utah, are closely associated with intrusive igneous rocks. Larger, more important deposits are found in essentially flat-lying limestone (Chap. 5) deposited in sedimentary basins in the inner parts of the continents far from igneous sources. These deposits, known as Mississippi Valley-type deposits, are thought to have been formed by hydrothermal waters circulating through the sedimentary basins after deposition of the limestone. Still others, such as the enormous Sullivan mine in British Columbia (Fig. 8-1) and the Broken Hill and Mount Isa deposits in eastern Australia, seem to be sedimentary deposits, possibly associated with hot spring activity, that formed about 1.3 to 1.5 billion years ago in seas in which little volcanic activity was occurring.

About 250 million tons of zinc and 140 million tons of lead are estimated by the U.S. Geological Survey to remain in presently known, minable deposits throughout the world, and several times this amount is thought to await discovery and exploitation at prices above today's range of 15 to 40 cents per pound. These are large numbers when compared to present annual mining rates of 6 million tons of zinc and about 3 million tons of lead. For zinc, it seems likely that consumption will continue to grow. Zinc has three main uses: zinc alloys for die casting, galvanizing, and brass, in that order of importance. Only in die-casting alloys is there a real possibility of substitution, largely by more expensive magnesium and aluminum.

For lead, the future is not so certain. There seems no doubt that it will continue to be used extensively in applications such as storage batteries, pipes, and lead sheeting, which account for over 50 percent of lead consumption. The uncertainty, however, lies in the use of about 3 grams of tetraethyllead, $Pb(C_2H_5)_4$, to increase the energy yield of each gallon of gasoline. This lead enters the environment via the engine exhaust and is lost completely from the recyclable lead inventory. Because most pipes and batteries are recycled, the lead in these antiknock compounds accounts for about one-third of annual world consumption of newly mined lead.

Concern over possible loss of the gasoline additive market for lead rests largely in the fact that excess lead is very harmful to humans, especially children. There is a lack of agreement on the human body's tolerance for low levels of lead, and particularly on what constitutes an unacceptable level of atmospheric lead. Even the available data, however, have convinced many scientists that worldwide environmental lead levels are rapidly approaching levels that could be harmful to large segments of our population. Available data indicate that today's humans contain more lead in their bodies than did their prehistoric counterparts and that lead is more abundant in all parts of today's environment (Table 8-2). Annual snowlayers in the Greenland ice cap, for instance, show that atmospheric lead fallout over Greenland has increased by more than 300 percent since 1940. Although some of this lead comes from combustion of coal, which contains about 25 ppm of lead, it is generally agreed that much of it comes from leaded gasoline, which contains 1,000 ppm lead. In response to this concern, many countries have begun to remove lead from their gasoline. Although this will undoubtedly minimize environmental lead levels, it is a two-edged sword in that removal of lead from gasoline will probably result in higher-priced, lower-octane gasoline which will, in turn, result in a more rapid depletion of our oil supplies.

MERCURY

Mercury, the new pariah of the mineral industry, makes lead look like penicillin. Only a few years ago, demand for mercury was skyrocketing, its price exceeded $700 a flask (76 pounds of mercury), and widespread concern was expressed about future mercury supplies. Those were the "good old days" for the mercury industry before demand slackened and the price halved in response to a growing concern about mercury pollution. This concern reached a peak in 1972 when the U.S. government banned pesticides, fungicides, and paints containing toxic mercury compounds (representing about one-quarter of U.S. mercury consumption) largely because they were also toxic to humans. The other 75 percent of U.S. mercury consumption is used largely in the elemental form, some of which is not particularly toxic. Proof of this is the fact that even with teeth full of mercury-rich dental amalgam, your hair is still not falling out, at least not from mercury poisoning.

Elemental mercury can cause problems, especially when released in vapor form, and miners in mercury deposits are subject to mercury poisoning from this source. The greatest problems are caused by mercury when it occurs as methylmercury (CH_3Hg^+) and related compounds, which are very toxic and readily taken up by organisms. Where such man-made "organomercurials" are dumped into the environment, as at Minamata, Japan, major catastrophies have resulted. Where simple elemental mercury enters the environment, such as from chlorine-caustic soda plants, the bacterially induced formation of organomercurials is slow, and similarly serious pollution problems have yet to be documented. In addition to these industrial sources of elemental mercury, some regions, such as the "barrens" west of Hudson Bay, exhibit anomalously high mercury contents that could not have resulted from industrial activity. The relative contributions of widespread natural mercury accumulations and localized industrial mercury sources to the rise in worldwide mercury levels remains to be established.

Table 8-2 Lead content* of air, soil, and plants at varying distances from a busy highway. Combustion of leaded gas produces lead particles, which settle out of the air rapidly thereby concentrating near roadways.

	DISTANCE FROM HIGHWAY (feet)		
Material	30	120	520
Air	2.3	1.7	1.1
Soil	65.0	40.0	25.0
Tomatoes	1.3	1.2	1.6
Beans	1.9	1.2	0.9
Potatoes	0.48	0.64	0.40

*Lead contents in micrograms per cubic meter for air and in micrograms per gram (ppm) for other materials.

SOURCE: Ter Haar (1970).

Against this background of uncertainty, we must try to assess future mercury supplies. Mercury deposits usually consist of veins of cinnabar (Appendix I) sometimes mixed with native mercury. Very few large deposits are known, and the largest, Almaden in Spain, has dominated world mercury production for 1,000 years. This circumstance reflects the relatively small size of world mercury consumption as well as the large size of the Almaden deposit. All important mercury deposits are found in geologically young mountain chains (Fig. 8-2). Recent estimates indicate that about 250,000 tons of mercury are minable from known deposits at the presently generous price of $5 a pound. Even larger price increases and discovery of new deposits are thought to have the potential to quintuple this figure. In relation to the present worldwide annual mercury consumption of 7,000 to 10,000 tons, these reserves are only moderate.

TIN

Tin is an element with friends. As new technological developments gradually displace it from its older uses, such as corrosion-resistant coatings on steel cans, other applications are found to offset these market losses. These new applications come largely from research funded by the major tin-producing nations. These nations, along with many important tin-consuming nations, form the International Tin Council (Table 8-3), which regulates the supply and price of tin. The ITC has attempted to encourage tin consumption by stabilizing tin prices. Despite these efforts, the price of tin almost doubled, between 1972 and 1974, to about $4 a pound.

Tin deposits usually consist of veins and disseminations of the mineral cassiterite (Appendix I) in or near granitic intrusive and volcanic rocks. Because cassiterite is more resistant to weathering and decomposition than its enclosing rocks, it often forms placer deposits (Chap. 10). Sizable amounts of tin are recoverable as by-products of some base metal mines,

Table 8-3 Producing and consuming members of the International Tin Council, a group which attempts to regulate the world supply and price of tin by production quotas and by buying and selling tin from its own stockpile. The United States is not a member because of objections to such international price-control groups.

PRODUCING NATIONS	CONSUMING NATIONS	
Australia	Austria	Italy
Bolivia	Belgium	Japan
Indonesia	Bulgaria	Korea
Malaysia	Canada	Luxembourg
Nigeria	Czechoslovakia	Netherlands
Thailand	Denmark	Poland
Zaire	Fed. Rep. Germany	Spain
	France	USSR
	Hungary	UK
	India	Yugoslavia

notably the Sullivan mine in British Columbia and the Kidd Creek mine in Ontario. Economic tin deposits are known in only a few countries (Fig. 8-2), and the outlook for discovery of major new areas is not encouraging. Minable world tin reserves are estimated to be over 3 million tons of metal with at least ten times as much thought to await discovery and development as price and demand escalate. These amounts are only moderate in relation to present annual consumption rates of nearly 200,000 tons.

TUNGSTEN

Tungsten metal, which costs about $4 to $7 a pound, is of value for its extremely high melting point of about 6100°F and its strength. When combined with other elements such as iron and carbon, it produces alloys of exceptional hardness, even at high temperatures. Although molybdenum, titanium, and aluminum can be substituted for tungsten in some applications, the results are usually inferior. The average abundance of tungsten in the earth's crust is .001 percent, and tungsten ore bodies must contain about 0.4 percent tungsten to be minable. Tungsten deposits usually consist of large veins or closely spaced smaller veins and disseminations of wolframite and scheelite (Appendix I), which were deposited by hot water related to nearby intrusive igneous rocks. The most important tungsten deposits are in geologically young mountain belts (Fig. 8-2). Over 60 percent of minable world tungsten reserves of about 2 million tons (of tungsten) are in southeastern China. In relation to recent world consumption of 20 to 30 thousand tons of tungsten, world reserves are comfortably adequate, provided they can reach world markets.

SMELTING

With the exception of tin and tungsten and nickel in laterites, the elements discussed here are combined with sulfur in their ores (Appendix I). After the ores are mined and the ore minerals concentrated from the worthless rock, the metal in the ore mineral must be separated from the sulfur by smelting. Since ancient times, smelting has been accomplished by pyrometallurgy, which involves heating the concentrate to convert the sulfur in the mineral to sulfur dioxide (SO_2) gas, leaving the molten metal behind. Smelting has become the focus of considerable criticism recently because it, along with the burning of fossil fuels, represents a major source of atmospheric SO_2 (Table 5-3). Whereas present smelting recovers as little as 20 percent of the sulfur from its emissions, some recently passed laws require that as much as 90 percent of the sulfur be recovered in the near future, a restriction considered too stringent by many smelting companies. The SO_2 recovery problem centers on the fact that although smelting emits large total amounts of SO_2, the concentration of SO_2 in smelter gases varies with time and is very low during part of the smelting process. Existing technology can clean the high SO_2 gases, but there are no proven methods to clean the low SO_2 gases economically.

Smelting also emits other gases, as well as metal-rich dust. Exploration geochemists have long known that soil and vegetation samples (Chap. 2) collected downwind from smelters are anomalously high in metals being recovered by the smelter. Recent studies have shown that dust and fumes released during lead and mercury smelting can yield dangerously high metal concentrations in air, soil, plants, and even people very near the operations.

Numerous efforts are being made to eliminate these problems. For one thing, new methods of pyrometallurgy are being developed to produce gases that are rich enough in SO_2 to be more easily cleaned. Techniques are also being perfected for nearly complete recovery of dust and other gases from smelter emissions. The even more radical approach of hydrometallurgy is also being investigated. This approach involves leaching the ore concentrate with a caustic solution and subsequent removal of the dissolved metal from solution. In many of these developing schemes, the caustic solution is recycled, there are no gaseous or particulate emissions, and the minimum economic size of the operation need not be as large as that of conventional smelters.

The enormous cost of existing smelters usually makes it impossible to abandon them immediately in favor of new designs in spite of the desirability of limiting SO_2 emissions. Thus, if metal demand prevails, it seems likely that we will see a relatively rapid, but not abrupt, decrease in SO_2 emissions within the next few years, probably accompanied by shortages in smelter capacity and rising metal prices.

CONCLUSIONS

Throughout this chapter we have encountered estimates on the order of 30 to 150 years for future supplies of the scarce metals. The validity of these estimates, of course, depends on hazy prognostications of reserves and future demand. Nevertheless, barring unforeseen breakthroughs in methods of exploration and ore production, it seems unlikely that they could be wrong by more than a factor of two or three in either direction. It seems, then, that we are living in a period of plenty that our not-too-distant descendants will not have. We must begin to recycle our metals more carefully to prolong man's time as a tool user.

NINE

MONEY AND DECORATIONS—PRECIOUS METALS AND GEMS

INTRODUCTION
Enough of this fertilizer, iron, and oil business! Let's shift attention to something everybody wants. And to get it, you would seem to need only a shovel and maybe a mule to help haul back your fortune. Or perhaps you would prefer a 747. The gems and noble metals discussed in this chapter have such a high value-to-weight ratio that they can be, and often are, transported by air. In fact, it is this very portability that has allowed the gems and precious metals to retain their special place in society. In times of war or economic uncertainty, a family can carry its fortune in these commodities literally in its pockets. Jewels, of course, are best suited for this practice, but the metals are also widely used. For instance, before the recent dramatic increase in the price of gold, it was estimated that over 550 million ounces of the metal were being hoarded.

Because of their value, the gems and precious metals in ornaments and money are given special care and probably represent man's closest approach to total recycling. Nevertheless, growing world population and affluence, as well as an ever-increasing number of industrial applications, have created a strong, continuing demand for further supplies of these mineral commodities. Let's begin with gems and work our way toward gold, gradually becoming more enmeshed in the monetary applications of the precious metals.

GEMS

Gems represent the ultimate in value for mineral resources. A three-carat diamond, which weighs only .0013 pound, retails for $1,000 to $10,000 depending on quality. It is no accident that diamonds, emeralds, rubies, and sapphires are the most expensive gems. They are among the hardest and therefore most durable of natural materials, and they reflect and refract light with unusual brilliance and beauty. In addition, they are rare. In the "rich" diamond mines of South Africa, for instance, gem-quality diamonds make up only about .000005 percent of the rock. Of the other 200 or so natural minerals, rocks, and other substances that have been used as gems, only a few, notably opal, pearls, and jade (Table 9-1), have received widespread acceptance.

There are very few minable concentrations of most gems. For instance, most of the world's high-quality emeralds come from a few small mines in the Andes of Colombia. Emeralds were found there before the discovery of the New World. The extreme rarity and small size of most gem deposits, as well as the small world demand for most gems, means that gem mining can be carried out by individuals and small groups. In the United States, for instance, most of the annual gem production, valued at between $2 million and $3 million, can be accredited to "rock hounds." This unusual situation develops partly from the fact that it is difficult to free gems from their enclosing host rock without fracturing them. However, many gems remain unchanged as the surrounding rocks weather to form soil, and can be recovered simply

Table 9-1 The most widely used gems. Diamond, emerald, ruby, sapphire, and chrysoberyl are usually the most expensive, whereas pearls, opal, and jade can be only moderately expensive.

GEM	CHEMICAL COMPOSITION	MAJOR SOURCE
Diamond	Carbon (C)	South Africa
Emerald	Beryllium aluminum silicate ($Be_3Al_2Si_6O_{18}$)	Colombia (Chivar and Muzo areas)
Ruby	Aluminum oxide (Al_2O_3)	Burma (Moguk area)
Sapphire	Aluminum oxide (Al_2O_3)	Ceylon (Ratnapura area)
Chrysoberyl (Alexandrite)	Beryllium aluminum oxide ($BeAl_2O_4$)	USSR (Sverdlosk) Ceylon
Opal	Silicon dioxide plus water ($SiO_2 \cdot nH_2O$)	Mexico (Queretaro) Australia (Coober Pedy) Burma (Uru area)
Jade	Jadeite (Sodium aluminum silicate) ($NaAlSi_2O_6$) Nephrite (Calcium magnesium silicate)	New Zealand, British Columbia, Alaska
Pearl	Calcium carbonate ($CaCO_3$)	Persian Gulf

by washing through the soil and stream gravels around known gem occurrences. Most of the world's supply of fine Burmese rubies is produced in just this way by a few hundred men.

Of all the precious stones, only diamonds are produced in sizable amounts. Annual world consumption of gem diamonds is over 12 million carats (about 3 tons) with raw (uncut) value of more than $600 million. Although gem-quality diamonds have been found throughout the world, 90 percent come from Africa and over 50 percent from South Africa (Fig. 9-1). Diamond, which is simply carbon in a very dense crystal form, is thought to have formed under high pressures at depths of over 100 kilometers in the

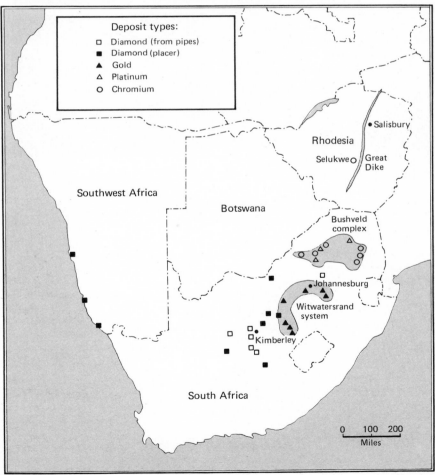

Figure 9-1 Generalized geologic map of southern Africa showing the location of important deposits of precious metals, diamonds, and chromium.

Figure 9-2 Aerial photograph of a diamond mine outside Kimberley, South Africa showing the steep sides of the pit indicating a grossly pipelike form for the ore. A natural diamond (still attached to its host rock) from one of the South African mines is shown in the inset. A penny provides scale for the diamond. *(Courtesy of S. B. Lumbers.)*

upper mantle of the earth. The South African diamond deposits are found in pipelike bodies made up of fragments of unusual, magnesium-rich rock called kimberlite, which appears to have been intruded upward from the mantle (Fig. 9-2). Diamonds are dispersed throughout some of these pipes and are mined largely by underground methods.

During the millenia before the diamond-bearing pipes were discovered, erosion of their surface exposures yielded diamonds that were concentrated in stream gravels and even coastal beaches as placer deposits (Chap. 10). Gem diamonds constitute only about one-third or less of the total diamonds in many mines in the kimberlite pipes, whereas many of the gravel and beach deposits contain almost nothing but gem diamonds. It is thought that the poorer quality, industrial diamonds (Chap. 4) were pulverized during stream transport.

Tremendous control is exercised over the diamond industry by de Beers Consolidated Mines, Ltd., of South Africa. This organization mines or markets (by agreement with the producers) over 80 percent of world gem diamond production, including that of the USSR, the number two producer.

This situation has allowed the price of gem diamonds to remain considerably above the level it would probably seek in a completely open market.

PLATINUM

It was not until 1844 that the silvery, hard nuggets being discarded from placer gold deposits in Colombia and the Urals were known to consist mainly of the elements platinum and palladium with smaller amounts of iridium, osmium, rhodium, and ruthenium. World production of the platinum group, as these elements are known, is 4 to 5 million ounces* annually, with about equal amounts of platinum and palladium accounting for over 90 percent of the total. Jewelry and ornaments consume only about 10 percent of production. The two largest uses are in electronic circuitry, where contacts must remain uncorroded, and as catalysts which, though not consumed in the chemical reactions, greatly facilitate production of nitric and sulfuric acids and petroleum refining.

Platinum and palladium producers may benefit in two ways from efforts to reduce automobile exhaust emissions. First, if it is necessary to eliminate lead from high-octane gasoline (Chap. 8), more platinum and palladium will be needed as catalysts to produce high-octane unleaded gas. Secondly, one of the best presently available systems for cleaning exhaust gases of nitrous oxides and hydrocarbons uses a platinum-palladium catalyst, which cannot function with leaded gasoline because the catalyst becomes coated with lead. This system, if used, would require about 500,000 ounces of platinum and 200,000 ounces of palladium annually. Because platinum costs over $200 an ounce and palladium costs about half as much, efforts are being made to substitute palladium for platinum and, wherever possible, to develop substitutes for both metals.

South Africa and the USSR produce over half of the world's platinum and palladium. Canada is third in both metals. South African production comes largely from the Bushveld complex, which we discussed in Chapter 7 as a major source of chrome ore (Fig. 9-1). In one of the chrome-rich layers, known as the Merensky Reef (Fig. 9-3), there are numerous platinum alloys and complex platinum sulfide minerals, which are mined largely by underground methods. The Merensky Reef extends along the surface for over 200 kilometers and provides large reserves. Russian and Canadian production comes from the nickel sulfide ores of Noril'sk (Fig. 8-2) and Sudbury (Chap. 8), respectively, where minor amounts of the platinum metals are combined with bismuth and arsenic. Reserves here depend largely on nickel reserves, which are large.

SILVER

For the last 15 years, the world has consumed far more silver than has been mined, leaving an annual deficit of 100 to 500 million ounces to be made up from other sources. Only about 20 percent of this deficit is filled by recycled silver. So, where does all the rest come from? Some silver is recovered from

---

*Precious metal abundances are measured in troy ounces. One troy ounce equals 31.1 grams or 0.0686 pound.

Figure 9-3 Underground mining of the Merensky Reef in the Bafokeng mine, South Africa. The platinum-rich part of the reef ranges in thickness from a few inches to over 10 feet, although most minable zones are relatively thin, as in this photograph. Over 25 minerals have been identified in the reef. The most abundant of these are pyroxene and plagioclase, of course, with important minor amounts of chromite. Minerals containing platinum-group metals include sperrylite ($PtAs_2$), laurite (($Rn,Os,Ir)S_2$), and braggite (($Pt,Pd,Ni)S$). *(From* Engineering and Mining Journal.*)*

coins. Coinage, however, once accounting for over half of world consumption, is no longer a major use for silver. The reason for this, in large part, is that an increase in the price of industrial silver can make the amount of the metal in a silver coin worth more than the face value of the coin. When this happens, the coins are melted or "mined" for their silver content. In the United States, this last happened in 1963 when the price of silver rose above $1.2929 an ounce. Between 1967 and 1970, the United States government sold 212 million ounces of silver obtained from old dimes and quarters. In 1968, such worldwide demonetization of coins supplied 40 percent of the deficit, but this source supplies less than 10 percent at present.

The bulk of the deficit, then, has been filled by sales of silver from stockpiles and hoards. Governments and industries with such supplies have contributed steadily to demand in recent years. Other silver, held by speculators in Europe and North America, has been released as silver prices have risen from about $2 to over $4 per ounce in the last few years. The largest silver stocks, however, are in India, where it is estimated that between 3.8 and 5.6 billion ounces of silver are hoarded. As a result of export restrictions in effect until recently, this silver entered world markets largely via the Persian Gulf sheikdom of Dubai, to which it was smuggled. Although Dubai has no silver mines, it "produced" about 30 million ounces of silver in 1973. Much greater amounts of silver are expected to leave India now that prices have increased and export restrictions have been relaxed.

Unfortunately, silver mine production cannot be increased greatly to fill the deficit because about 75 percent of new silver production is a by-product, or coproduct, of other metal mining. Therefore, demand for these other metals controls silver production rates. Most of the major metal deposits we discussed in Chapter 8 supply by-product silver. Sudbury and other Canadian nickel sulfide deposits produce over 1 million ounces annually, as does the White Pine copper deposit in Michigan. Porphyry copper deposits produce 8 million ounces in the United States alone. Lead-zinc deposits, from Broken Hill, Australia to the Mississippi Valley deposits of the United States, produce silver. Perhaps the most impressive by-product silver source is the Kidd Creek massive sulfide (copper-zinc) deposit in Ontario, which, with a production of 12 to 14 million ounces a year is, by accident, the world's largest "silver mine."

Most deposits in which silver is mined as the major element are vein systems where the silver minerals were deposited by circulating hydrothermal solutions (Chap. 8). In the Comstock lode in Nevada and the many similar deposits in Mexico and the Western United States, the veins extend downward only about 600 meters, and their formation was closely associated with volcanic activity. In the Coeur d'Alene area of Idaho (Fig. 8-2), which has produced over 750 million ounces of silver and large amounts of lead and zinc, the veins extend to depths of 2,500 meters and were probably formed by waters related to intrusive igneous rocks. In almost all silver mines, the silver is found in galena, in silver sulfide, or in complex silver-arsenic-antimony sulfide minerals (Appendix I).

GOLD

Gold is a loner. It does not usually combine with other elements but commonly occurs in nature as rich, yellow native gold that resists tarnish and corrosion. Much native gold is easily visible in and recoverable from its ores, and it can be fashioned into useful and decorative shapes. Thus, two-thirds of noncommunist industrial gold consumption, about 45 to 50 million ounces annually, is used in jewelry and the arts. The remainder goes largely into electronic circuitry and dentistry. Substitutes for gold are not generally available, but many applications, such as your class ring, are not really essential.

The widespread use of gold in jewelry led to its use as a medium of exchange, first as ornaments, then as coins, and finally as government gold stocks to validate paper money. Over 40 percent of the 3 billion ounces of gold estimated to have been mined during recorded history is in government and bank vaults for this purpose. As international trade became the mainstay of world prosperity, it was deemed necessary to establish an official world price for gold on which all transactions could be based. At the Bretton Woods conference in 1944, the United States, the most powerful economic force in the world, undertook this task by agreeing to set the value of gold at $35 an ounce for all its domestic and foreign transactions and to use its large government gold stocks to maintain world prices at this level.

The rapid increase in world population and affluence, however, required that large amounts of new money be printed to maintain an adequate per capita monetary supply. But with the price of gold fixed at $35 an ounce and mining costs steadily rising, world mines could not supply enough gold to validate this new currency. In fact, world gold mine production, which has fluctuated since 1933 in the range of 25 to 50 million ounces annually, has even fallen short of industrial demand on many occasions. In the United States, much of the domestic gold demand had to be supplied by sale from the government gold stocks that were needed to stabilize the world gold price.

Sensing that a likely solution to this supply problem would be an increase in gold prices, foreign governments exchanged dollars for gold. Speculators also began to hoard gold, which further aggravated the supply problem. Almost half of 1968 world gold production, for instance, went into private speculative holdings. By this time, the U.S. gold reserve had dropped to less than half its 1949 total of about 70 million ounces, and it was no longer practical to maintain world gold prices at $35 an ounce. Thus, through a complex and still evolving series of devaluations, revaluations, and government policy changes, industrial gold was set free to find its own market level, now near $150 an ounce, and the gold price for intergovernment transactions was last set at $42.22 an ounce.

One of the main beneficiaries of these events has been South Africa, where over two-thirds of the world's gold is produced from the 2.2-billion-year-old Witwatersrand sedimentary basin (Fig. 9-1), commonly called the "Rand." This basin consists largely of sands and gravels that have been hardened into rock through time and which contain small grains of gold and uraninite (Chap. 6). The Rand ores appear to be colossal analogues of the largely exhausted placer gold deposits along the west flank of the Sierra Nevada mountains in California, which caused the great gold rush of 1849. Whereas the placers of the Sierra Nevada eventually led to their source in the gold vein deposits of the Mother Lode, the Rand deposits have no obvious source. Nevertheless, they have yielded almost one-third of the world's gold.

Other large gold deposits, such as the Homestake mine in South Dakota and the Kerr-Addison mine in Ontario, could be Precambrian gold-rich sedimentary layers that were later broken and reconstituted by circulating hydrothermal solutions. Gold is also found in veins formed during volcanic activity, as in the now-exhausted Comstock and Goldfield areas in Nevada, El Oro in Mexico, and the recently discovered Pueblo Viejo deposit in the Dominican Republic. Finally, in some of the newly discovered deposits in Nevada, such as Carlin, the gold is so fine-grained that it is invisible even in that time-honored prospecting tool, the gold pan. From deposits such as these, and by-product gold largely from copper mining, the United States, the USSR, Canada, and Australia supply most of the remainder of world production.

South Africa holds about 60 percent of the 1 billion ounces of gold estimated to be minable throughout the world. At prices of about $150 an ounce, this gold represents a formidable economic resource. This total amount is only a 10- to 30-year supply, and without South African reserves, the supply is almost nothing. Clearly, we have not seen the last of interesting developments concerning gold.

TEN

WHAT CAN WE GET FROM TODAY'S OCEANS?

INTRODUCTION
Throughout this book, we have concerned ourselves largely with mineral resources that are available on land. Many geologists feel that this is only part of the picture and that the oceans, too, will contribute to our growing mineral demands. After all, the oceans cover almost three-quarters of the globe but are only sketchily explored. In evaluating the mineral potential of the ocean, we will divide it into two parts, ocean water and the ocean floor, each of which is discussed in a short section. As it turns out, there are some potentially interesting mineral resources in the ocean, which raises the dual problems of how to recover the materials and what sort of legal system governs their recovery. These questions, which really form the crux of the ocean resource problem, merit a short closing section.

OCEAN WATER
Can you quell the romance of the sea long enough to critically examine the composition of ocean water (Table 3-2)? If so, you can see right away that this brew is no gold mine. Remember that rocks with less than about 1 ppm gold are economic untouchables. So how can we expect to recover gold from

sea water, which has over 100,000 times *less* gold, even if it is easy to process? Of course, it sounds great to hear that each cubic mile of ocean water contains 40 pounds of gold, but the old problem of grade is still with us, and traces of metal such as this are not likely to be recoverable in the near future.

Without the rose-tinted glasses, you might correctly conclude that the highly concentrated elements in sea water, such as sodium, chlorine, and magnesium (Table 3-2), would be the most promising economically. In fact, almost half of total world magnesium production is obtained by a complex chemical process in which roasted oyster shells are mixed with sea water to precipitate magnesium hydroxide. Over five percent of U.S. salt and an even larger percentage of world salt (Chap. 5) is produced by evaporating sea water, using solar heat, in large shallow ponds. Bromine can also be recovered economically from sea water. Even to our voracious industrial appetite, the ocean is essentially infinite and there is therefore no concern about our future supplies of these commodities.

THE OCEAN FLOOR

Configuration of the Ocean Floor Most of the ocean floor is either shallowly submerged or very deeply submerged. The shallow parts of the ocean floor are actually submerged extensions of the continents in most areas and are known as the continental shelves. The average water depth on the continental shelves is about 300 feet, though many areas such as the Grand Banks of Newfoundland are shallower. Over 80 percent of the ocean floor, however, is 6,000 to 20,000 feet deep, a long way to reach. These regions, which constitute the deep-ocean floor, are much less easily observed and sampled and have remained largely unknown until recently. In addition to its remoteness, the deep-ocean floor is underlain by oceanic crust, which is chemically and geologically distinct from the continental crust that underlies the continental shelves.

Continental Shelves Because the continental shelf is the submarine extension of the continental crust, anything we find on the continents should be found in rocks of the shelves, also. Miners have known this for a long time and have followed good ore zones in Alaska, England, Japan, and other areas for many miles beyond shore by underground mining (Fig. 10-1). Probes such as this, though locally impressive, are insignificant with respect to the 11 million square miles of the world's shelves. Thus, we must confine our present economic interest in the continental shelves to mineral resources we can dredge off the surface of the shelf or those we can obtain from the rocks underlying the shelf by drilling and pumping.

Petroleum and natural gas (Chap. 6) are by far the most important mineral resources derived from drilling and pumping in the continental shelves. In recent years, almost 20 percent of world petroleum production has been from offshore wells (Fig. 10-1). Sulfur, which can be melted by hot water pumped down a well, can also be recovered from beneath the continental shelf as can soluble mineral resources such as potash that might be extracted by solution mining (Chap. 5).

The continental shelf is covered by a veneer of shells, silt, sand, and gravel that contains two main classes of economically interesting mineral resources. Most obvious are the sands, gravels, and shells themselves, which are used largely as construction materials in areas such as metropolitan New York where suitable resources on land are either not available or exhausted (Fig. 10-1). Of more interest to us here, however, are placer deposits, in which there is an unusual concentration of valuable minerals (Table 10-1) that are relatively dense and resistant to chemical decomposition during weathering. These minerals tend to concentrate where water currents are slowed slightly, and they form placer deposits at bends in streams or in the zone of surf action on a beach (Fig. 10-2). So why should we look for them on the continental shelves? The reason, of course, is that during the last major glacial period much of the world's water was locked up on land in the large ice sheets causing sea level to be much lower. For instance, about 15,000 years ago sea level was over 300 feet lower than it is now, and streams and rivers flowed out across the shelf. Thus, almost anywhere there are interesting placer deposits on land, we can expect to find their extensions on the adjacent submerged shelf. This line of reasoning has led to the discovery of placer gold off Nome, Alaska and placer tin mines in the Sunda shelf around Indonesia (Fig. 10-1). The most important of these, so far, are the tin placers of coastal Thailand which have been in production for over 70 years and which supply over 5 percent of present world tin needs. Such placer deposits

Table 10-1 Minerals commonly found in marine placer deposits. Note that all these minerals have a high specific gravity in relation to common minerals, such as quartz and feldspar (specific gravity 2.5 to 3.0) and water (1.0). The extremely high specific gravity of gold accounts in part for its common occurrence in placer deposits. Compositions of these minerals are given in Appendix I.

MINERAL	MATERIAL RECOVERED	APPROXIMATE SPECIFIC GRAVITY
Cassiterite	Tin	7
Chromite	Chromium	4.5
Columbite-tantalite	Niobium (columbium) and tantalum	5–8
Diamond	Diamond	3.5
Gold	Gold	19.5
Ilmenite	Titanium dioxide	5
Magnetite	Iron	5
Monazite	Thorium and rare earth elements	5
Platinum	Platinum	17
Rutile	Titanium dioxide	4
Scheelite	Tungsten	6
Wolframite	Tungsten	7
Zircon	Zirconium-hafnium	5

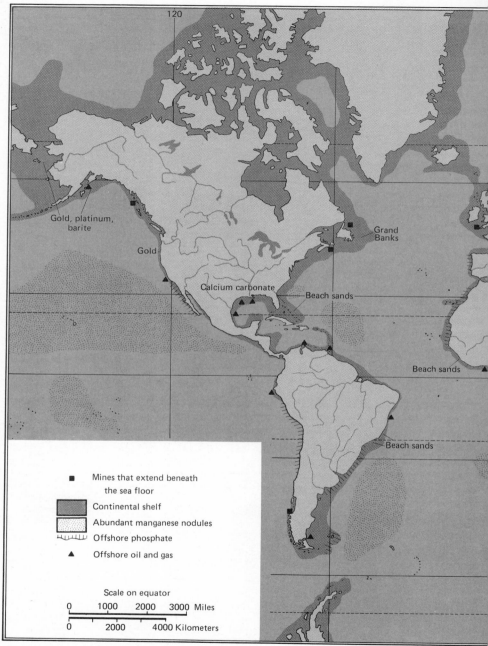

Figure 10-1 Distribution of continental shelf and deep-ocean basins showing the location of mineral resources of present and potential future economic importance. *(After Mero, 1965; Wenk, 1969; Anderson, 1972; and Horn, 1972)*

104 OUR FINITE MINERAL RESOURCES

Figure 10-2 These beach sediments, rich in rutile, ilmenite (Chap. 4), and zircon, are mined by dredging along the east coast of Australia. The small, floating structure at the left cuts its way toward the left pumping dredged sediment to the larger, floating structure, which is a concentrator. Waste from the process is sprayed out beside and behind the concentrator gradually filling the lake behind the dredge. *(From* Engineering and Mining Journal.*)*

of the continental shelf have been mined only where they are relatively accessible, and the outlook is good for the discovery of additional deposits.

One other submarine mineral resource of the continental shelves merits attention simply from the standpoint of its abundance. Remember the Phosphoria formation of the Western United States that we described in Chapter 5 as the source of so much of the world's phosphate fertilizers? This formation was formed in an ancient ocean, as you probably recall, and generally similar deposits are forming today on some of the continental shelves (Fig. 10-1). These submarine phosphate deposits are largely of lower grade than the older, more easily mined deposits on land, however, and thus have not yet been exploited. They constitute a reassuring reserve of this important element.

Deep-Ocean Floor The rapidly increasing fund of information on the deep-ocean floor has provided several startling revelations to geologists and geophysicists in the last few decades. The most exciting, of course, is the fact that the sea floor is spreading outward from certain zones at rates of about 0.2 to 2 inches per year and thereby causing the continents to drift about on the surface of the globe. Sampling of the sea floor has also disclosed the presence of manganese nodules and metal-rich muds, both of which have excited considerable economic interest.

Manganese nodules are black, spherical to irregular concretions of manganese and iron oxides, usually a few inches in diameter, which are amazingly abundant on the deep-ocean floor (Fig. 10-3). An estimated 1.5 trillion tons of nodules lie on the sea floor of the Pacific Ocean, alone. Some areas contain an estimated 300,000 tons of nodules per square mile, and more commonly observed concentrations, on the order of 30,000 to 75,000 tons per square mile, are considered economically interesting. In addition to their high manganese and iron contents (Table 10-2) the nodules contain important but variable amounts of copper, nickel, cobalt, zinc, lead, molybdenum, and zirconium. All available evidence indicates that the nodules are growing continually by addition of dissolved manganese and other elements from sea water. Growth rates for individual nodules are almost negligibly small, but there are so many of them that the aggregate addition of elements with time is relatively large. The purest nodule concentrations are developed where no sediments are being introduced to dilute the chemical growth of the nodules, which explains their development in the deep-ocean floor, remote from sources of land-derived sediment (Fig. 10-1).

Chemical tests indicate that nodules recovered from the ocean can be crushed and processed, probably by hydrometallurgy (Chap. 8), to separate the minor elements such as copper and nickel. The remaining material is then a manganese ore very similar in grade and chemistry to that obtained from terrestrial manganese mines. Thus, considerable effort has been put

Figure 10-3 Manganese nodules, such as these from a depth of 5,320 meters in the central south Pacific, cover much of the deep-ocean floor. In some areas, these nodules are also found below the sediment surface. The area in this photograph measures approximately 7 feet from side to side, making individual nodules about 3 to 5 inches in diameter. *(From The Seafloor Photograph Collection, Lamont-Doherty Geological Observatory, Palisades, New York.)*

Table 10-2 Average composition of manganese nodules from different parts of the Pacific Ocean. All these nodules came from the surface of the ocean floor. (Some areas also contain buried nodules.) Note the marked variation in minor element content of the nodules. This factor will probably be as important as abundance in selection of areas suitable for mining.

	NORTHEAST PACIFIC	SOUTHEAST PACIFIC	CENTRAL PACIFIC	SOUTH PACIFIC	NORTH PACIFIC
Mn (%)	22.33	19.81	15.71	16.61	12.29
Fe (%)	9.44	10.20	9.06	13.92	12.00
Ni (ppm)	10,800	9,610	9,560	4,330	4,220
Co (ppm)	1,920	1,640	2,130	5,950	1,440
Cu (ppm)	6,270	3,110	7,110	1,850	2,940
Pb (ppm)	280	300	490	730	150
Mo (ppm)	470	370	410	350	180
Ti (ppm)	4,250	4,670	5,610	10,007	6,340

SOURCE: Cronan (1972).

into development of a workable process for mining the nodules, a problem to be discussed in the next section.

The metal-rich muds are somewhat more localized, although we have not yet looked for them in all the right areas. The best known of this type is in the Red Sea at a depth of about 6,000 feet (Fig. 10-1). Here hot, dense, metal-rich brines and mud have filled several depressions in the sea floor. The upper 30 feet of the deposit alone is estimated to contain 50 million tons of this material but as much as 80 percent of it is actually brine. Typical metal concentrations, in both wet and dried samples of the deposit (Table 10-3), are not sufficient to encourage mining efforts at present. Nevertheless, there could be other similar, but higher-grade deposits elsewhere on the ocean floor. The extensive copper-iron sulfide deposits of Cyprus are a possible example of the results of an ancient ore-forming process similar to

Table 10-3 Composition of metal-rich mud from a shallow depression in the floor of the Red Sea at a depth of about 6,000 feet. The mud, when originally collected, contains about 60 percent brine (analysis A) and therefore has a much lower metal content than the oven-dried mud (analysis B). The material in its natural state is water-saturated, and analysis A is a truer representation of its grade *in situ*.

	A WET ASSAY (60% brine)	B DRY ASSAY
Copper (percent)	0.45%	1.1%
Zinc (percent)	0.90%	2.3%
Silver (ounces per ton)	2.90 oz	7.25 oz
Gold (ounces per ton)	0.02 oz	0.05 oz

SOURCE: Walthier and Schatz (1969).

that presently active in the Red Sea. In Cyprus, the ore is found as sulfide lenses in rock thought to be old oceanic crust that was later thrust upward to form an island.

As was pointed out earlier, the oceanic crust, which underlies the deep-ocean floor, is quite different from the continental crust. The difference lies in the fact that the deep-ocean floor is compositionally distinct from the continents and, in almost all areas, has always been deeply submerged and has not undergone many of the processes that have formed ore on the continents. Evaporite deposits and conventional deposits of lead, tin, uranium, lithium, and many other elements are almost certainly absent from the deep-ocean floor. About the most important aspect of this difference is the general lack of petroleum in the deep-ocean basins. One of the few exceptions to this rule is the recent discovery of hydrocarbon indications in a well, drilled for scientific purposes, in the deep, central part of the Gulf of Mexico. Petroleum accumulations in the deep Gulf of Mexico had been suspected previously by some geologists on the basis of salt dome-like structures outlined in the area by seismic surveys. The origin and nature of these possible hydrocarbon accumulations are not known, although it is likely that they resulted from an unusual sequence of events, possibly involving the sliding of continental shelf sediment into the deep Gulf. Whatever the reason, it seems unlikely that there will be significant petroleum production from the deep-ocean floor.

PROBLEMS OF OCEAN MINING

As our brief summary has shown, there is no doubt about the abundance of economically interesting mineral resources on both the continental shelf and the deep-ocean floor. In relation to what is there, however, production of resources other than oil and gas is very small. This is because the minerals of the land areas have always been available by standard engineering methods and under fairly uniform laws. As yet, there are no equivalent methods or legal systems to stimulate ocean-floor production, nor any assurance that their development would make such production competitive in cost.

By far the greatest progress toward efficient engineering methods for exploiting the ocean floor has been made by the petroleum industry. Offshore oil and gas drilling and pumping equipment consisting of platforms supported by legs anchored to the sea floor are commonplace in many areas. These units can stand and produce in water up to 300 feet deep, and designs are available for $10 million units that will work in over 600 feet of water. For deeper water, the industry has developed floating platforms and ships capable of working in water up to 2,000 feet deep and costing up to $30 million each. Much of this equipment was designed for use in the North Sea where it must be capable of operating in 30-foot waves, and of surviving 90-foot waves (Fig. 10-4).

In contrast, most equipment being used at present for exploration and production of solid submarine mineral resources is relatively primitive, although interesting developments have occurred recently. Ocean-bottom mining is commonly done by suction dredge, by bucket dredge (using an

Figure 10-4 Western Oceanic Pacesetter I is a $20 million semisubmersible offshore drilling rig. This unit averaged 9.18 knots in its 5,366-mile trip from Beaumont, Texas to the North Sea using its own power and one tug. When traveling, Pacesetter I has its derrick in a horizontal position and has a draft of 17 feet. When drilling, its submerged pontoons are partly flooded to give it a draft of 60 feet and a wave clearance of 50 feet. The unit weighs almost 20,000 tons when in service. Note the helicopter pad at left. *(From* Oil and Gas Journal.*)*

endless belt equipped with scoops), or by a grabbing device called a clamshell. Most dredges include not only the suction or digging device, but also the mineral beneficiation equipment on the same vessel or an accompanying barge. These are among the cheapest forms of mining equipment, but they cannot operate efficiently in rough water. This is a special disadvantage because many actively forming ocean placer deposits are in shallow water or in the surf zone where navigation is treacherous. Such conditions caused the closing of dredging enterprises for gold near Nome and for diamonds in southern Africa (Fig. 9-1). (The African diamond deposits are so rich, reaching about $30 per ton versus more common grades of $1 to $5 per ton in other placer deposits, that some deposits are mined by building dikes into the ocean and pumping out the water to permit conventional mining.) Another problem with ocean dredging has been that of lifting large volumes of material from great depths. Because of these problems most interest in ocean dredging has been limited to depths of less than 150 feet outside the surf zone.

Interesting progress has been made lately on both of these engineering fronts. For work in the surf zone, both submersible and bottom-attached dredges have been developed. Much more important, however, has been the development of dredges that can reach the deep-ocean floor. Two systems have been tested successfully. One, a suction dredge, has been operated at a depth of about 2,000 feet. The other, an ingenious continuous bucket dredge, has performed satisfactorily at depths of over 12,000 feet. Both of these systems appear to be capable of mining at a rate of about 1,000 tons or more per day, a typical rate of operation for a medium-sized, high-grade mine.

As the hardware for deep-ocean pumping and mining has developed, a new problem has arisen. Who owns deep-ocean floor deposits? This question is the major restraint in deep-ocean mining. Why invest millions in a mining system that you might never be able to use economically? The most widely accepted international legal convention governing the ocean at present recognizes state control over the water and mineral resources up to 3 miles offshore (though many countries claim areas much farther from shore) and state control of the mineral resources out onto the continental shelf to a depth of about 600 feet. Beyond this limit, where many deep shelf deposits and all deep-ocean floor deposits lie, ownership has not been adequately defined. Efforts to revise these regulations to accommodate the growing quantities of known, deep-ocean mineral resources is of concern to the United Nations at present. Most proposals to the United Nations have the common element of declaring areas over 200 miles offshore or beyond the 600-foot depth limit to be international sea floor, to be mined in such a way that all countries, even those without sea front acreage, share in the profits. This could be accomplished if the United Nations granted exploration and exploitation permits and then charged a royalty on any production, but the nature of the royalty must not discourage operations.

A final element of concern in considerations of the feasibility of ocean mining is the environmental question. Although most older stream and beach dredgers were not members of the garden club, present-day dredging, like strip mining (Chap. 6), can be controlled to minimize damage and prevent unsightliness. The concern over ecological imbalances that might result from deep-ocean mining also seems to have abated recently as it has become apparent that the amount of particulate material and nutrient-rich deep water transferred to the surface will be negligible.

WILL WE EVER REALLY MINE THE OCEANS?

If the economic incentive is clarified, extensive exploitation of our ocean mineral resources may become a reality in a relatively short time. In spite of this optimistic assessment, we should not lose sight of the fact that the oceans will increase reserves of most mineral resources only fractionally. For petroleum and natural gas, for instance, we can hardly expect that the continental shelf contains more reserves than does the land surface, which occupies three times as much area. Using an optimistic approach, nodule mining might be expected to multiply our land-based resources of manganese, copper, and nickel by several times, though a lower figure seems more likely for the near future. Thus, although we can expect the oceans to alleviate impending shortages, they will not provide us with inexhaustible resources.

ELEVEN
READ THIS

We have by no means completed the list of marketable mineral commodities. There are the "glamor metals" such as beryllium and niobium that are used in special structural applications, and also tantalum, which is widely used in electronic equipment. Elements such as zirconium and hafnium are closely linked to the development of nuclear power in that zirconium metal will not limit the intensity of fission reactions whereas hafnium is a very effective moderator of these reactions. Strontium, a rather common element, provides the red color of many fireworks and ammunition and also shields color television viewers from harmful picture tube radiation. Other elements, such as scandium, which is used in high-intensity lamps, are just finding applications in today's industrial society.

For these elements, as well as most of those discussed in this book, however, the outlook is about the same—spotty distribution of recoverable or even potentially recoverable reserves and limited future resources. What can we do? What will happen as mineral consumption increases and resources decrease? In a world untouched by war, plague, government regulation, and birth control, the most likely assumption is that consumption will continue to rise until demand equals resource availability, whereupon scarcities will develop and consumption will slowly decline as resources dwindle. This theoretical pattern has received widespread attention and is generally accepted. Many worried citizens, basing their concern on this sort of analysis, have pointed out quite convincingly that such an obviously finite supply of mineral resources leaves little room for doubt about the ability of the earth to sustain continued exponential growth of mineral demand. They note further that this, in turn, will limit the earth's population and/or industrial capacity within the next few centuries. These warnings are serious and merit careful attention by all. We must decide to what extent our actions as consumers reflect the value of the finite resources we are exhausting.

At the same time, and perhaps more importantly, we must not forget that the "end" is still in the future. There is time to act, and what we do now may govern the condition of our descendants' lives. As indicated in Chapter 1, the ultimate magnitude of our mineral resources depends as much on economic, political, and technological developments as on geology. Much of the ore mined today was considered waste 10 years ago. Breakthroughs in copper exploration, aluminum production, natural gas transport, and coal gasification will continue to transform today's waste into tomorrow's ore.

This is the optimistic view and the one that encourages exploration and development of our mineral resources while at the same time promoting sensible consumption. This is where you as a citizen come into the picture. If we are to ensure continuing adequate mineral supplies, we must all take as active an interest in the legislation and government regulations affecting our mineral resources as we do in foreign relations and domestic welfare. Did you know, for instance, that the governments of Mexico and Manitoba spend millions of dollars each year in mineral exploration, presumably on behalf of their citizens? Is this wise economy when private industry is willing to do the same thing if given an opportunity to make a profit? Do you agree that the oil shales of the Western United States should be developed now to supply U.S. energy needs, or should they be set aside as an energy resource for the future? Over 80 percent of the oil shales are on public land, and a 5,120-acre tract was recently leased to private oil firms for over $210 million. Was that too little or too much? Should Canada continue to export over 1 million barrels of Alberta oil to the United States each day, while importing a larger amount from Venezuela and the Middle East to supply the eastern provinces?

Similar questions can be posed for all mineral commodities and for all of them the basic problem remains the same. Should a country attempt to maintain self-sufficiency in its dwindling mineral resources or should it import mineral supplies and run the risk of depending on unstable foreign sources? Should it export its mineral products, thereby gaining revenues but hastening the day when the resources are depleted? Questions like these are what motivate increasingly louder calls for a "national mineral policy" in most countries. Such mineral policies must be formulated by qualified teams including geologists, engineers, and economists. Too often, however, a national mineral policy is a patchwork of hastily prepared, politically inspired legislation, all too frequently demanded by an uninformed public that simply wants the problem to vanish, if only for another year.

It's no exaggeration to say that a national mineral policy (or the lack of one) constitutes the base on which industrial pyramids are built and maintained, particularly in the face of growing mineral shortages. At this point in history, there are too many of us at the top of the pyramid looking toward the gathering gloom of future mineral shortages and too few of us at the foundation intelligently influencing and formulating the policies that will govern mineral availability over the next century.

APPENDIX ONE

IMPORTANT ORE MINERALS

ORE MINERALS FROM WHICH SPECIFIC ELEMENTS ARE OBTAINED

ELEMENT OBTAINED	SYMBOL	ORE MINERALS	COMPOSITION	MAXIMUM CONTENT OF DESIRED ELEMENT
Aluminum	Al	Bauxite		
		Boehmite	$Al_2O_3 \cdot H_2O$	43
		Diaspore	$Al_2O_3 \cdot H_2O$	43
		Gibbsite	$Al_2O_3 \cdot H_2O$	34
Antimony	Sb	Stibnite	Sb_2S_3	72
		Valentinite	Sb_2O_3	83
		Tetrahedrite	$Cu_8Sb_2S_7$	25
Arsenic	As	Arsenopyrite	FeAsS	46
		Löllingite	$FeAs_2$	73
		Smaltite	$CoAs_2$	72
Barium	Ba	Barite	$BaSO_4$	59
Beryllium	Be	Beryl	$Be_3Al_2Si_6O_{18}$	5
		Bertrandite	$Be_4Si_2O_7(OH)_2$	15
Bismuth	Bi	Native bismuth	Bi	100
		Bismuthinite	Bi_2S_3	81
Boron	B	Borax	$Na_2B_4O_7 \cdot 10H_2O$	12
Cadmium	Cd	Sphalerite	$(Zn,Cd)S$	<1
Cesium	Cs	Pollucite	$(Cs,Na)AlSi_2O_6$	43
Chromium	Cr	Chromite	$(Mg,Fe)(Cr,Al,Fe)_2O_4$	44
Cobalt	Co	Cobaltite	$(Co,Fe)\,AsS$	35
		Pyrrhotite	$(Fe,Ni,Co)_{1-x}S_x$	<1
Copper	Cu	Native copper	Cu	100
		Chalcopyrite	$CuFeS_2$	35
		Chalcocite	Cu_2S	80
		Bornite	Cu_5FeS_4	63
		Enargite	Cu_3AsS_4	48
Fluorine	F	Fluorite	CaF_2	49
		Cryolite	Na_3AlF_6	54

APPENDIX ONE

ELEMENT OBTAINED	SYMBOL	ORE MINERALS	COMPOSITION	MAXIMUM CONTENT OF DESIRED ELEMENT
Gold	Au	Native gold	Au	100
		Calaverite	$AuTe_2$	44
Iron	Fe	Magnetite	Fe_3O_4	72
		Hematite	Fe_2O_3	70
		Limonite	$Fe_2O_3 \cdot H_2O$	60
		Siderite	$FeCO_3$	48
		Pyrite	FeS_2	47
Lead	Pb	Galena	PbS	87
Lithium	Li	Spodumene	$LiAlSi_2O_6$	8
		Lepidolite	$KLi_2AlSi_4O_{10}F_2$	8
		Petalite	$LiAlSi_4O_{10}$	5
Magnesium	Mg	Magnesite	$MgCO_3$	28
Manganese	Mn	Pyrolusite	MnO_2	63
		Psilomelane	$BaMn_9O_{18} \cdot 2H_2O$	52
Mercury	Hg	Liquid mercury	Hg	100
		Cinnabar	HgS	86
Molybdenum	Mo	Molybdenite	MoS_2	60
Nickel	Ni	Pentlandite	$(Fe,Ni)_9S_8$	36
		Pyrrhotite	$(Fe,Ni,Co)_{1-x}S_x$	<2
Niobium (Columbium)–Tantalum	Nb-Ta	Columbite-Tantalite	$(Fe,Mn)(Nb,Ta)_2O_6$	66(71)
Platinum group				
Platinum	Pt	Sperrylite	$PtAs_2$	57
Palladium	Pd	Froodite	$PdBi_2$	20
Iridium	Ir	Laurite	$(Ru,Ir,Os)S_2$	61(75,75)
Osmium	Os			
Rhodium	Rh			
Ruthenium	Ru			
Rare-earth metals				
Yttrium	Y	Monazite	$(Ce,La,Th,Y) PO_4$	—
Important lanthanide		Bastnäsite	$CeFCO_3$	64
elements		Xenotime	YPO_4	48
Cerium	Ce			
Europium	Eu			
Lanthanum	La			
Rhenium	Re	Molybdenite	$(Mo,Re)S_2$	<1
Selenium	Se	Copper ores	—	—
Silver	Ag	Native silver	Ag	100
		Argentite	Ag_2S	87
		Argentiferous galena	$(Pb,Ag,Bi,Sb)S$	—
		Argentiferous tennantite—tetrahedrite	$(Cu,Fe,Ag)_{12}(As,Sb)_4S_{13}$	—
Strontium	Sr	Celestite	$SrSO_4$	48
Sulfur	S	Native sulfur	S	100
		Gypsum	$CaSO_4$	24
		Pyrite	FeS_2	47
Tellurium	Te	Copper ores	—	—
Thorium	Th	Monazite	$(Ce,La,Th,Y) PO_4$	71
Tin	Sn	Cassiterite	SnO_2	79
Titanium	Ti	Rutile	TiO_2	61
		Ilmenite	$FeTiO_3$	32
Tungsten	W	Wolframite	$(Fe,Mn)WO_4$	76
		Scheelite	$CaWO_4$	80
Uranium	U	Uraninite	UO_2	88
		Coffinite	$U(SiO_4)_{1-x}(OH)_{4x}$	60
		Carnotite	$K_2(UO_2)_2(VO_4)_2 \cdot 3H_2O$	57
Vanadium	V	Carnotite	$K_2(UO_2)_2(VO_4)_2 \cdot 3H_2O$	12
Zinc	Zn	Sphalerite	ZnS	67
		Willemite	Zn_2SiO_4	56
Zirconium and hafnium	Zr(Hf)	Zircon	$(Zr,Hf)SiO_4$	49(<2)

ORE MINERALS COMMONLY USED IN MINERAL FORM

ORE MINERAL	COMPOSITION	MAJOR USES
Anhydrite (see gypsum)	—	—
Asbestos		
Chrysolite	$Mg_6Si_4O_{10}(OH)_8$	Filler in cement and plaster (Chap. 4)
Barite	$BaSO_4$	Used in oil well drilling to "float" rock cuttings and prevent blowouts; also in glass (Chap. 4)
Boron minerals		Glass and chemical industries (Chap. 4)
Borax	$Na_2B_4O_7 \cdot 10H_2O$	
Kernite	$Na_2B_4O_7 \cdot 4H_2O$	
Calcite (limestone, marble)	$CaCO_3$	Fillers (Chap. 4); production of lime (CaO) and other calcium chemicals (Chap. 5).
Chromite	$(Mg,Fe)(Cr,Al,Fe)_2O_4$	Refractory compounds (Chap. 7)
Clays		Construction materials, fillers, carriers, and extenders (Chap. 4)
Kaolinite	$Al_4Si_4O(OH)_8$	
Montmorillonite (fuller's earth)	$Al_2Si_4O(OH)_2 \cdot xH_2O$	
Diamond	C	Abrasive (Chap. 4)
Diatomite (the microscopic shell of an aquatic organism)	$SiO_2 \cdot nH_2O$	Filler (Chap. 4); also used extensively to filter liquids
Feldspar		Ceramics, abrasive, filler (Chap. 4)
Microcline	$KAlSi_3O_8$	
Plagioclase	$(Na,Ca)Al_2Si_2O_8$	
Fluorite	CaF_2	Blast furnace flux.
Graphite	C	Refractories, lubricant, electrical equipment
Gypsum (anhydrite)	$CaSO_4 \cdot 2H_2O$; $(CaSO_4)$	Plaster, cement (Chap. 4, 5)
Kyanite	Al_2SiO_5	Refractory compounds
Limestone (see calcite)	—	
Magnesite	$MgCO_3$	Refractory compounds
Mica		Filler (Chap. 4), electrical insulation
Muscovite	$KAl_2(AlSi_3O_{10})(OH)_2$	
Phlogopite	$KMg_3(AlSi_3O_{10})(OH)_2$	
Vermiculite	$Mg_3Si_4O_{10}(OH) \cdot xH_2O$	Thermal insulation, packaging
Ocher, umber, sienna	Iron oxide with variable amounts of silica, alumina, manganese oxide, and water	Pigment (Chap. 4)
Phosphate minerals		Fertilizer, phosphorus chemicals (Chap. 5)
Apatite	$Ca_5(PO_4)_3(F,Cl,OH)$	
Potassium minerals		Fertilizer, potassium chemicals (Chap. 5)
Sylvite	KCl	
Carnallite	$KCl \cdot MgCl_2 \cdot 6H_2O$	
Pyrophyllite	$Al_2Si_4O_{10}(OH)_2$	Filler (Chap. 4)
Silica sand		Glass (Chap. 4)
Quartz	SiO_2	
Sodium minerals		Sodium chemicals (Chap. 5)
Halite (salt)	$NaCl$	
Trona	$Na_2CO_3 \cdot NaHCO_3 \cdot 2H_2O$	
Nahcolite	$NaHCO_3$	
Talc	$Mg_3Si_4O_{10}(OH)_2$	Filler (Chap. 4)
Titanium dioxide minerals		Pigment (Chap. 4)
Rutile	TiO_2	
Ilmenite	$FeTiO_3$	
Wollastonite	$CaSiO_3$	Filler (Chap. 4)
Zeolites	A large family of complex hydrous sodium, calcium, aluminium silicates	Ion-exchange medium for water purification, etc.

TWO APPENDIX

PRODUCTION, SOURCES, AND PRICES OF MINERAL RESOURCES OF IMPORTANCE IN WORLD TRADE.

ELEMENT OR COMMODITY	ESTIMATED WORLD MINE PRODUCTION (1973)	MAJOR ORE SOURCES	APPROXIMATE PRICE RANGE* (1973–1974)
Aluminum	61,900,000 tons (aluminum ore of all types) (1972 production)	Australia Jamaica Surinam USSR Guyana France	$0.29 to $0.39/pound (aluminum metal)
Antimony	78,650 tons	South Africa Bolivia Mexico USSR	$0.55 to $1.75/pound (antimony metal)
Asbestos	4,422,000 tons	Canada USSR South Africa	$35 to $1,750/ton (depending on fiber-length, etc.)
Barite	4,373,000 tons	USA W. Germany Mexico Peru USSR	$15 to $80/ton (depending on purity, etc.)
Bismuth	4,300 metric tons	Peru Bolivia Australia Mexico USA	$4 to $8/pound (bismuth metal)

ELEMENT OR COMMODITY	ESTIMATED WORLD MINE PRODUCTION (1973)	MAJOR ORE SOURCES	APPROXIMATE PRICE RANGE* (1973–1974)
Cadmium	22,000 tons (est.)	USA Japan USSR Canada	$3 to $4/pound (cadmium metal)
Chromium and chromite	1,299,000 tons (1971)	Rhodesia South Africa USSR Turkey	~$1.50/pound for chromium metal and $33–$40 dollars/long ton for chromite
Coal	3,328,436,000 tons (1971 production, all types of coal)	USSR USA China E. Germany W. Germany Poland	$5 to $20 per ton
Cobalt	24,093 metric tons	Zäire Zambia Canada	$2 to $3/pound (cobalt metal)
Copper	7,500,000 tons	USA Canada Zambia Chile Zäire	$0.50 to $1.30/pound (copper metal)
Fluorspar (fluorine)	5,150,291 tons (year ending June 30, 1973)	Mexico Spain Thailand France Italy	$60 to $100/ton of fluorite (depending on purity, etc.)
Gold	43,400,000 troy ounces	South Africa USSR Canada USA Australia	$60 to $200/troy ounce (gold metal)
Gypsum	63,500,000 tons	USA Canada France USSR	almost all gypsum is sold in processed or fabricated commodities such as plaster or wallboard
Iron and steel	858,000,000 metric tons (iron ore); 763,000,000 tons (raw steel)	USA Canada Australia Brazil (iron-ore producers)	$11 to $13/long ton (iron ore); 8 to 10 cents/pound (finished steel)
Lead	2,942,000 tons (excluding USSR and China)	USA Australia Canada Mexico Peru	$0.14 to $0.30/pound (lead metal)
Lithium	9,400,000 pounds of lithium equivalent (consumption)	USA Rhodesia	$8 to $9/pound (lithium metal)—most is sold as lithium chemicals and minerals.
Magnesium	200,000 tons (largely from brines and sea water)	USA Norway USSR	$0.40 to $0.50/pound (magnesium metal)
Manganese	22,420,000 tons (manganese ore)	USSR South Africa Gabon Brazil Australia	$0.03 to $0.06/pound (48% Mn ore)
Mercury	270,200 flasks (76 pounds each) of mercury	Spain Italy USSR Mexico	$250 to $300/flask (76 pounds of mercury)

APPENDIX TWO

ELEMENT OR COMMODITY	ESTIMATED WORLD MINE PRODUCTION (1973)	MAJOR ORE SOURCES	APPROXIMATE PRICE RANGE* (1973–1974)
Molybdenum	151,000,000 pounds of Mo in concentrate (excluding USSR and China)	USA Canada Chile	$1.80 to $2.20/pound of contained Mo in molybdenum compounds
Natural gas	37,907 billion cubic feet (1970 production)	USA USSR Canada Netherlands Romania Mexico	$0.15 to $0.80/1000 cubic feet
Nickel	1,000,000,000 pounds	Canada New Caledonia USSR Cuba Australia	$1.54 to $1.62/pound (nickel metal)
Niobium (columbium)	17,000,000 pounds (est.)	Brazil Canada Nigeria	$2.45 to $3.05/pound of Nb in ferro-columbium alloy
Petroleum	17 billion barrels (est.)	USA Saudi Arabia Iran Venezuela Kuwait Libya Nigeria	$3 to $15/barrel (crude oil)
Phosphate	108,700,000 tons of "phosphate rock"	USA USSR Morocco Tunisia Togo	$10 to $15/ton of "phosphate rock" (65–75% P_2O_5)
Platinum group	2,700,000 troy ounces (platinum); 2,200,000 troy ounces (palladium)	South Africa USSR	Pt—$160 to $250/ounce; Pd—$60 to $100/ounce
Potash	22,000,000 tons K_2O equivalent	Canada USA E. Germany W. Germany USSR	$39 to $75/ton of ore containing 62% K_2O equivalent
Rare-earth elements	26,000 tons of rare-earth oxides	USA Australia India Malaysia	$4 to $440/pound pure oxide (depending on element)
Selenium	2,500,000 pounds (from smelting of copper ores)	USA Canada Japan	$10 to $18/pound (selenium metal)
Silver	249,000,000 troy ounces	Canada Peru Mexico USA	$2 to $6/troy ounce (silver bullion)
Sulfur	33,750,000 long tons sulfur equivalent (excluding USSR and China)	USA Canada Poland France USSR	$7 to $31/ton (pure sulfur)
Tantalum	2,000,000 pounds tantalum equivalent (excluding USSR)	Brazil Canada Nigeria Zaire	$30 to $65/pound (tantalum metal)

ELEMENT OR COMMODITY	ESTIMATED WORLD MINE PRODUCTION (1973)	MAJOR ORE SOURCES	APPROXIMATE PRICE RANGE* (1973–1974)
Tellurium	460,000 pounds (from smelting of copper ores)	USA Canada Peru Japan	$6 to $7/pound (tellurium metal)
Thorium	15,500 tons of monazite concentrate (see Appen. I.)	Australia India Brazil Malaysia	$99 to $113/ton of monazite concentrate
Tin	192,700 metric tons	Malaysia Bolivia Thailand Indonesia	$1.79 to $3.51/pound (tin metal)
Titanium	29,000,000 pounds titanium metal (U.S. production)	USA USSR Japan	$1.18 to $1.42/pound (titanium sponge)
Tungsten	44,600,000 pounds (excluding USSR and domestic Chinese production)	China Canada Australia USA	$4 to $7/pound (tungsten metal)
Uranium	25,000 tons U_3O_8 in ore (est.)	USA Canada South Africa France	$6 to $10/pound ($U_3O_8$)
Vanadium	38,000,000 pounds (consumption)	USA South Africa	$3 to $4/pound of vanadium in ferro-vanadium alloy
Zinc	6,275,000 tons	Canada USA Australia Peru Japan	$0.20 to $0.40/pound (zinc metal)
Zirconium	7,500,000 pounds (consumption)	USA	$9 to $26/pound (zirconium metal)

*The price range reflects variations in purity of the commodity as well as variations in price during this period.

INDEX

Abrasives, 39–40
Aggregate, 31–32, 34, 38
Almaden mercury mine, 87
Aluminum, 1, 3, 6, 7, 74–75, 77, 86, 89
Antimony, 78, 97
Arsenic, 78, 95, 97
Asbestos, 34–36, 38, 41
Athabasca tar sands, 61
Autlan manganese mines, 76

Bathurst lead-zinc-copper mines, 17, 21
Bauxite, 74–75
Beneficiation, 5, 21, 54
Beryllium, 110
Bingham copper mine, 79, 85
Bismuth, 78, 95
Blast furnace, 32, 67, 73
Boron, 37
Brines, 37, 47, 106
Broken Hill lead-zinc mines, 86, 97
Bromine, 110
Bushveld complex chrome-platinum mines, 76, 95
By-products, 84, 88, 97, 98

Cadmium, 78
Carlin gold mine, 98
Carnotite, 62
Cassiterite, 88
Chalcopyrite, 79, 81
Charco Redondo manganese mines, 76
Chiatura manganese mines, 76
Chromium, 9, 57, 67, 76–77, 95
Cinnabar, 88
Clay, 36, 38–39, 69, 74–75
Climax molybdenum mine, 84
Coal, 3, 21, 45, 54–56, 67–68, 86
Cobalt, 78, 145
Coeur d'Alene lead-zinc-silver mines, 97
Coke, 66–67
Comstock lode, 23–24, 97, 98
Concrete, 34–35, 38
Continental shelf, 100–104
Copper, 3, 5, 6, 9, 14–17, 21, 24, 74, 78–84, 98, 105, 109
Copper Belt copper mines, 81
Cornwall iron deposits, 79
Corundum, 39

Demonitization (of silver), 96
Deuterium, 63
Devaluation, 98
Diamonds, 7, 39–40, 92–95, 108

Dimension stone, 32–33
Dolomite, 39, 48
Drilling, 18–19
(See also Mining methods, wells)

Economics (of mineral deposits), 4–12, 46, 79–80
El Oro gold mines, 98
El Teniente copper mine, 9
Emeralds, 92–93
Emery, 39
Energy, 1, 52–65, 74
Engineering (aspects of mineral deposits), 4, 19–21
Environmental problems, 19, 29, 34, 40–41, 50–51, 57–59, 63, 73–74, 84, 86–87, 89–90, 109
Ertsberg copper mine, 6
Evaporites, 36, 37, 42–48, 63, 81, 107
Exhaustion (of mineral resources), 2, 4, 8, 27–28, 61, 65, 69, 110–111
Exploration (for mineral resources), 4, 6, 8, 12–19, 56–57, 81–82, 110–111
Extenders, 39
Extraterrestrial mineral resources, 6–7

Feldspar, 36–37
Fertilizers, 47–51
Fillers, 38–39
Fission energy, 61–63
Fluorine, 67, 74, 77
Fossil fuels, 53–61
Fuller's earth, 39
Fusion energy, 61–64

Galena, 85, 97
Garnet, 39
Gasoline, 56–57, 86, 95
Gems, 91–95
Geochemical exploration, 15–17, 89–90
Geophysical exploration, 17–18, 109
Geothermal power, 64–65
Glass, 36–37
Gold, 4–6, 14, 21, 62, 95, 97–101, 108
Goldfield gold mines, 98
Government actions (related to mineral resources), 8, 9, 13–14, 40, 47, 54, 73, 79, 97–98, 109, 111
Grade, 5, 6, 45, 62, 63, 66, 77, 79, 81, 84, 104, 105, 108
Granite, 33

Great Dike chromite deposits, 76
Green River saline lake deposits, 36, 61
Groundwater, 21, 24–28, 45–48, 50–51, 54, 64, 81
Gypsum, 34–36, 44–45, 47

Hafnium, 110
Hammersly Range iron mines, 68
Heavy oil, 60
Helium, 57
Hematite, 68
Henderson molybdenum mine, 84
Homestake gold mine, 14, 98
Horne (Noranda) copper-zinc mine, 81
Hubbert, M. K., 3, 61, 63
Hydrometallurgy, 90, 105
Hydrothermal solutions, 36, 39, 77, 79–80, 85, 89

Ilmenite, 38, 40, 104
Indium, 78, 95
Industrial minerals, 31–51
International Tin Council, 88
Iron, 6, 17, 18, 68–73, 81, 104–105
Isotope, 61n.

Jade, 92
Jeffrey asbestos mine. 41

Kalahari manganese mines, 76
Kambalda nickel mines, 24, 85
Kaolin, 36, 38–40
Kerr-Addison gold mine, 98
Kidd Creek copper-zinc mine, 14, 17, 81, 89, 97
Kiruna iron deposits, 73
Kupferschiefer copper mines, 81
Kuroko copper-zinc-lead deposits, 81

Labrador trough iron mines, 68
La Brea tar pits, 15, 56
Lake Superior iron ores, 68
Lakes, 29, 37, 48, 61
Laterite deposits, 74, 84, 89
Lead, 21, 77, 78, 81, 85–87, 90, 95, 105, 107
Lightweight aggregate, 32–34, 36
Limestone, 66
Limonite, 68
Lithium, 37, 40, 107
LNG (liquified natural gas), 57

INDEX

Magnesium, 37, 39, 86, 100
Magnetite, 39, 68
Manganese, 67, 75—76, 104—106
Marble, 33
Marmora iron mine, 14
Massive sulfide deposits, 79, 81—82, 85, 97, 106—107
Mercury, 78, 87—88, 90
Merensky Reef chrome-platinum deposits, 95
Metal-rich muds, 81, 104—107
Methane, 56, 57
Methylmercury, 87
Mica, 37, 39
Middle East oil fields, 57
Mineral deposit, defined, 4—12
Mining methods:
 general, 5, 19—21, 33, 45—47, 54, 61, 82, 84, 93, 95
 ocean mining, 100—101, 107—109
 solution mining, 47, 100—101
 strip mining, 54—56, 59, 60, 75
 wells, 5, 20—21, 27—28, 56—57, 60, 100—101, 107
Mississippi delta oil fields, 57
Mississippi Valley lead-zinc deposits, 86, 97
Moanda manganese mines, 76
Molybdenum, 78, 79, 84, 89, 105
Monazite, 62
Mother lode, 98
Mount Isa lead-zinc deposits, 86
Muscovite, 39

National minerals policy, 111—112
Natural gas, 5, 13—14, 18, 21, 37, 45, 48, 56—59, 100—101, 107, 109
Nepheline syenite, 37, 75
New Caledonia nickel mines, 84
Nickel, 23—24, 34, 78, 84—85, 95, 105, 109
Nickel sulfide deposits, 79, 85, 97
Niger delta oil fields, 57
Nikopol manganese mines, 85
Niobium, 110
Nitrate, 47
Nonrenewable resources, 2, 27—28
Noril'sk nickel mines, 95
North American Water and Power Alliance, 30
Nuclear power, 61—64, 110
Nuclear wastes, 63, 64

Ocean, mineral resources in, 99—109
Ocher, 37
Oil shales, 36, 60—61, 111

Opal, 92
Organomercurials, 87
Osmium, 95

Palladium, 95
Panarctic syndicate, 13—14
Panguna copper mine, 79
Parral lead-zinc mines, 77
Pearls, 92
Peat, 54
Pegmatites, 37, 39
Pellets (of iron ore), 69—73
Permafrost, 59
Petroleum, 2—6, 9, 13—15, 18, 19, 37, 45—48, 56—59, 74, 86, 100—101, 107, 109
Phosphate, 45, 47—51, 62, 104
Pigments, 8, 37—38
Pine Point lead-zinc mines, 17, 86
Pipelines, 54—57, 59
Placer deposits, 38, 93—94, 98, 101—104
Plaster, 35—36, 39, 44—45
Platinum, 95
Plutonium, 63
Porphyry copper deposits, 79—80, 97
Portland cement, 33, 34, 36, 44—45
Potash, 3, 6, 37, 47—48, 50—51, 100—101
Precambrian rocks, 54, 62, 68—69, 80, 81, 85
Prudhoe Bay oil field, 57
Pueblo Viejo gold mine, 98
Pumice, 33
Pyrite, 54, 68
Pyrometallurgy, 89

Quartz, 36, 39

Recovery (of mineral resources from mines and reservoirs), 47, 56
Recycling (of mineral resources), 3, 24, 86, 91, 95
Red Mountain iron deposits, 79
Refining (of oil and gas), 56—57, 95
Reserves, defined, 22
Reservoir, 27, 56
Resources, defined, 22
Rhenium, 78
Rubies, 92—93
Ruthenium, 95
Rutile, 38, 104

Saline lakes, 37, 61
Salt, 27, 45—47, 50, 63, 100, 107
Sand and gravel, 31, 32, 34, 101
Sapphires, 92
Scandium, 110
Scheelite, 89
Sea floor, 75, 104
Sea water, 24—26, 42—43, 47, 51, 63, 81, 99, 100
Sedimentary copper deposits, 79, 81

Sedimentary iron formation, 68—73
Selukwe chrome mines, 76
Serpentinite, 34—35
Siderite, 68
Sienna, 37
Sierrita copper mine, 79
Silver, 3, 6, 14, 21, 95—97
Slag, 32, 67
Smelting, 21, 50, 67, 89—90
Soda ash, 37
Sodium, 36—37, 100
Sphalerite, 5—6, 21, 85
Steel, 1, 3, 66—68, 75—77, 84
Strontium, 90—93, 110
Substitutions, 3—4, 79, 86, 89, 95
Sudbury nickel mines, 85, 95, 97
Sulfur, 7, 44—47, 49, 50, 54, 57, 60, 89—90, 100—101
Sullivan lead-zinc mine, 86, 89
Surface water, 26, 81, 73
Sylvite, 47
Synthetic hydrocarbon fuels, 59—61

Taconite, 69—73
Tailings, 22, 73
Talc, 39
Tantalum, 110
Tar sands, 59—60
Technological developments, 1, 2, 5, 7—8, 75, 85, 88, 90, 111
Tetraethyllead, 86, 95
Thallium, 78
Thompson nickel mines, 85
Thorium, 62—63
Tin, 78, 88—89, 101—104, 107
Titanium, 38, 39, 89
Transportation, 40—41, 44, 47, 54—55, 57—59, 68—69, 91
Tritium, 63
Trona, 36
Tungsten, 78, 89

Umber, 37
United Nations, 9, 109
U.S. Geological Survey, 3, 62, 63, 73, 86
Uraninite, 62, 98
Uranium, 1, 8, 17, 61—62, 98, 107

Vanadium, 3, 67
Vein, 62, 85, 97, 98

Water, 2, 23—30, 61
Weathering, 34, 36, 38, 69, 74, 76, 81, 84, 88, 92, 101
White Pine copper deposits, 81, 97
Witwatersrand basin, 62, 98
Wolframite, 89

Zinc, 5, 14, 17, 21, 37, 77, 78, 81, 85—86, 105
Zirconium, 105, 110